From Crust to Core

Carbon plays a fundamental role on Earth. It forms the chemical backbone for all essential organic molecules produced by living organisms. Carbon-based fuels supply most of society's energy, and atmospheric carbon dioxide has a huge impact on Earth's climate. This book provides a complete history of the emergence and development of the new interdisciplinary field of deep carbon science. It traces four centuries of history during which the inner workings of the dynamic Earth were discovered, and it documents the extraordinary scientific revolutions that changed our understanding of carbon on Earth forever: carbon's origin in exploding stars; the discovery of the internal heat source driving the Earth's carbon cycle; and the tectonic revolution. Written with an engaging narrative style and covering the scientific endeavors of about 150 pioneers of deep geoscience, this is a fascinating book for students and researchers working in Earth system science and deep carbon research.

SIMON MITTON is a life fellow at St. Edmund's College, University of Cambridge. For more than 50 years he has passionately engaged in bringing discoveries in astronomy and cosmology to the general public. He is a fellow of the Royal Historical Society, a former vice-president of the Royal Astronomical Society and a fellow of the Geological Society. The International Astronomical Union designated asteroid 4027 as Minor Planet Mitton in recognition of his extensive outreach activity and that of Dr. Jacqueline Mitton.

From Crust to Core

A Chronicle of Deep Carbon Science

SIMON MITTON

University of Cambridge

CAMBRIDGE
UNIVERSITY PRESS

CAMBRIDGE
UNIVERSITY PRESS

University Printing House, Cambridge CB2 8BS, United Kingdom

One Liberty Plaza, 20th Floor, New York, NY 10006, USA

477 Williamstown Road, Port Melbourne, VIC 3207, Australia

314–321, 3rd Floor, Plot 3, Splendor Forum, Jasola District Centre, New Delhi – 110025, India

79 Anson Road, #06–04/06, Singapore 079906

Cambridge University Press is part of the University of Cambridge.

It furthers the University's mission by disseminating knowledge in the pursuit of education, learning, and research at the highest international levels of excellence.

www.cambridge.org
Information on this title: www.cambridge.org/9781108426695
DOI: 10.1017/9781316997475

© Simon Mitton 2021

First published 2021

Printed in the United Kingdom by TJ Books Limited, Padstow Cornwall

A catalogue record for this publication is available from the British Library.

ISBN 978-1-108-42669-5 Hardback

For Marie Edmonds

Contents

Foreword

I came to know the curious, absorbent and learned Simon Mitton in
2012, when Simon completed the editing and publication of *Taking
the Back off the Watch: A Personal Memoir* of the scientist Thomas
Gold. As a graduate student of Nobel Prize-winning astronomer
Martin Ryle during a rather turbulent time and a historian of science
at the University of Cambridge, Simon knew Tommy, who was
conjecturing about the origin of very distant radio signals. Following
Tommy's death in 2004, Simon worked closely with Tommy's
widow, Carvel, to bring the memoir to fruition.

In 1995, grants from the Alfred P. Sloan Foundation supported
Gold in writing a book based on his path-breaking 1992 paper in the
Proceedings of the National Academy of Sciences paper, "The Deep,
Hot Biosphere." Gold's 1999 book, *The Deep Hot Biosphere: The
Myth of Fossil Fuels*, addressed controversial questions, including the
possibility that life originated deep in Earth rather than in a warm
little pond on its surface or extraterrestrially, arriving from space on a
comet or meteorite. It also argued that a large fraction of Earth's
hydrocarbons (natural gas, oil and coal) had primordial, abiotic origins
and accumulated in the crust from upward outgassing rather than
forming as "fossil fuels" from the shallow burial of biomass during the
Jurassic and other epochs.

While the Foundation took no position on Gold's propositions,
Sloan president Ralph Gomory (1990–2007), mathematician and
former chief of research for IBM, believed that big questions of the
kind Gold raised usefully stimulated science. Sloan also supported
Renegade Genius, a television documentary on Gold that appeared in
2009, five years after Gold had passed away.

In 2007, geologist Robert Hazen (Carnegie Institution for
Science, Washington, DC) published the book *Genesis*, aimed at

public understanding of questions associated with the origins of life. The book came to my attention as a Sloan program manager who had handled Sloan's grants with Gold. Hazen's book dispassionately weighed evidence for and against several of Gold's propositions, as well as pointing to other major unanswered questions in the geosciences, including the ecology and evolution of minerals.

Sloan invited and provided funds to Hazen to organize a May 2008 conference to explore the limits of knowledge (the known, unknown and unknowable) of deep carbon science, which Hazen did together with his colleague, Russell Hemley, then director of Carnegie's Geophysical Laboratory and an expert in materials under extremely high pressures, as in Earth's interior. In the interim, MIT microeconomist Paul Joskow had assumed the presidency of Sloan. President Joskow had a keen interest in energy resources and a long-standing relationship with Richard Meserve, president of the Carnegie Institution and former head of the US Nuclear Regulatory Commission. The May 2008 meeting proved very lively. President Joskow asked me to consult experts and stakeholders and to prepare an internal strategy paper for a 10-year international initiative about deep carbon science to be anchored by funding from Sloan and to convene an expert group to vet the strategy.

Among those consulted and immediately enthusiastic were earthquake expert, former Sloan Trustee and National Academy of Sciences president Frank Press and also Walter Munk (Scripps Institution of Oceanography), who helped start the international program for drilling in the seafloor during the 1960s, about which Simon writes in Chapters 8 and 9. Gold's former colleague John Saul (geologist) and student Steven Soter (astronomer) also provided advice and impetus. Favorable vetting resulted in an invitation in early 2009 from Sloan to Carnegie for a three-year US$4 million grant to initiate a major program in deep carbon science, spanning biology as well as solid-earth sciences. Hemley suggested the framework of a "Deep Carbon Observatory" (DCO) to emphasize the importance of making new measurements. The Sloan Trustees approved the grant in

June 2009, and the DCO officially came into existence on July 1, 2009, with Hazen as lead scientist and Hemley chairing a distinguished international steering committee. Founding members included John Baross (USA), Taras Bryndzia (Australia/USA), Claude Jaupart (France), Adrian Jones (UK), Barbara Sherwood Lollar (Canada), Eiji Ohtani (Japan) and Sergei Stishov (Russia). An administrative secretariat was established at Carnegie. Assuming the DCO proceeded very well, the Sloan Trustees committed to provide about US$5 million a year for 10 years, a total of about $50 million.

Over the next two years, a series of workshops led to the emergence of four communities to carry out the work of the DCO: Deep Life; Deep Energy; Reservoirs and Fluxes; and Extreme Physics and Chemistry. Each community prepared a set of "decadal goals" to achieve by the end of 2019 and agreed to organize the work along the four themes of origins, quantities, movements and forms. Cross-community teams concerned with data science and with engagement (communications and community building) were formed in 2011–2012. Early grant-making focused on instrument development, using both open, competitive calls for proposals and invited proposals. Subsequently, the majority of Sloan DCO funding was used to support a global network of graduate students and postdocs. Much effort went into community building: for example, cultivation of DCO support in Germany, France, Italy, the UK, Russia, China and Japan, as well as the USA; development of a website for both internal and external purposes; and giving the DCO a recognizable identity and family feeling. The program used the major international meetings in geosciences (especially the annual meeting of the American Geophysical Union and the annual Goldschmidt conference in geochemistry) to bring together the growing network of participants in the DCO.

At the outset of the program, Sloan asked the DCO to prepare a report that would describe the baseline of knowledge about deep carbon and that could be used to help measure progress achieved by 2019. The DCO leadership chose to try to create not only a benchmark, but a landmark, and in 2013, they published the 20-chapter, 698-page open access volume, *Carbon in Earth*. Released at an "all-program" meeting of close to 200 people at the US National Academy of Sciences in Washington, DC (March 2013), along with a press release summarizing the DCO's goals, the volume also served to attract many more scientists to the DCO network, which grew to about 500 by the end of 2013. A December 2013 press release highlighted early discoveries. The December 2014 Mid-term Scientific Report by Hemley summarized the first five years of the program. Subsequent all-program meetings took place in Munich (April 2015) and St. Andrews (March 2017).

A question early in the DCO decade was whether to foster an effort to drill through the crust into the mantle, as was strongly advocated by Japanese members. After the spring 2011 Tohoku earthquake, Japan deferred interest in this "Mohole."

During 2014, Sloan organized a far-reaching mid-term review by an external group of experts who had no stake in the DCO. The review led to major additions and changes in the program, including formation of a new group to take responsibility for synthesis, led by Cambridge volcanologist Professor Marie Edmonds, to whom Simon dedicates this book. The Synthesis Group and much of the strategic management of the DCO were handled by a team at the University of Rhode Island led by Sara Hickox and later Darlene Trew Crist. The Rhode Island team skillfully organized an October 2015 workshop that formulated most of the synthesis activities of the DCO. Creatively, they invited Simon to offer a historian's perspective on the DCO, which he did during a lively and provocative evening session. Sloan then invited Simon to submit a proposal to write a history of deep carbon science to place the DCO in context. The happy result is this book, which spans from the center of Earth to faraway habitable planets, with rich intervals in Europe during the Renaissance and Enlightenment, as well as the contemporary archipelagos of global research.

During 2015–2019, the active membership of the DCO network reached a total of about 1200 scientists from about 50 nations. In the later years, the substantial flow of DCO peer-reviewed publications included numerous papers in *Science, Nature* and other prestigious outlets. Under Edmonds and Trew Crist's leadership, a 50-page decadal report, press releases, special issues of journals and hundreds of other articles, as well as videos and blogs and two other books reachable at deepcarbon.net, summarize the work of the program. Although tracking matching and leveraged funds is difficult because of different forms of funding in different nations and for other reasons, a cautious estimate is that US$200–$250 million in funds from other sources complemented US$57 million that Sloan spent on the DCO program between 2009 and the culminating events in the fall of 2019.

I first met Tommy Gold in about 1983. In subsequent years, he would appear unexpectedly at my office at The Rockefeller University, having arrived on the Big Red Bus that shuttles Cornell University faculty and students between the main campus in rural Ithaca, New York, and the Manhattan campus. He would speak for an hour or so about abiotic methane, or the possible deep origin of life, or the formation of diamonds, and then abruptly depart. I believe Tommy was a Renegade Genius, and that even Tommy, who was schismatic, would have admired the contributions of the DCO, of which he was the progenitor. And he would especially have liked this book of Simon Mitton, which shows vividly how the matter of deep carbon has arrived as a science through the overthrow of received ideas.

Jesse H. Ausubel

Science Advisor to the Alfred P. Sloan Foundation for the Deep Carbon Observatory; Director, Program for the Human Environment, The Rockefeller University

Acknowledgments

The making of this history depended on the selfless support, encouragement, advice and assistance of a large number of colleagues, friends and family. I shall begin with my sincere thanks to members of the Department of Earth Sciences, University of Cambridge. I am deeply grateful to Marie Edmonds, who offered me this opportunity to contribute to the educational legacy of the *Deep Carbon Observatory* (DCO). When I undertook this challenge, I had less knowledge of the field than a beginning graduate student. Marie has been my tutor, director of studies, mentor, friend and enthusiastic supporter throughout. She never failed to answer an email by return. Whenever I asked her to read a draft chapter, I received feedback in a couple of days, always offering encouragement alongside correction. One of Marie's doctoral students, Fiona E. Iddon, was my editorial assistant, who read every chapter through three drafts, removing my goofs, duplications, complications and obfuscations, as well as imposing some discipline on my disorderly references. Fiona compiled the biographical notes and glossary in the back matter of this book. Penny Wieser expertly drew many of my line diagrams. Dan McKenzie, Simon Redfern and Sarah Humbert were unstinting in their support and encouragement. In the initial phase of the project, I benefitted from informative discussions with James Badro, Matthieu Galvez, Adrian Jones, Bénédicte Menez, Dominic Papineau, Graham Shields, Frederick Vine and Robert S. White.

At the Carnegie Institution for Science, home of the DCO, the Executive Director Robert Hazen gave advice in abundance: what to read, where to visit, whom to contact. Bob's boundless energy and his track record on writing books for a general readership were inspirational. In a similar context, I am hugely grateful to Steve Shirey, Douglas Rumble, Conel Alexander, Yingwei Fei, Shaun Hardy and the late Erik Hauri. I extend thanks also to Craig Schiffries, Craig Manning, Darlene Trew Crist, Andrea Mangum, Jennifer Mays, Michelle Hoon-Starr, Katie Pratt, Peter Fox and Josh Wood. At the Sloan Foundation, Jesse Ausubel was always there with his solid support, great enthusiasm and eager anticipation of the finished book.

I am most grateful to the following organizations for their support of my research over the past four years: St. Edmund's College, University of Cambridge, for acting as the grant administrator, and the American Geophysical Union and the European Geosciences Union for allowing access to the press facilities at their Meetings in 2016, 2017, 2018 and 2019. The kindness and professional engagement of their respective Press Officers in making me welcome and providing a good working environment are much appreciated.

My wife Jacqueline Mitton has given me enormous encouragement, advice and practical aid throughout the execution of this project, particularly during my extended period of medical leave when it seemed the book might never be finished. I am eternally grateful to her for her loving care and generosity.

The research and editorial costs of making this book were supported by the generous Book Grant G-2016-7283 awarded by the Alfred P. Sloan Foundation to St. Edmund's College, University of Cambridge. I am deeply grateful for this support.

Introduction

My engagement with the new field of deep carbon science began in September 2015, when Marie Edmonds of the Department of Earth Sciences at the University of Cambridge asked if I would be interested in researching a history of deep carbon science. By then I had been working in the history of science for some 15 years, following early retirement from Cambridge University Press. Much of my academic activity had been limited to the history of astronomy and cosmology in the twentieth century. In the geosciences, my knowledge of its pioneers was more or less limited to what had been achieved by people in Cambridge, particularly in tectonics, which was centre stage when I commenced my doctoral research at the Cavendish Laboratory in the radio astronomy group that had discovered pulsars. Over a working lunch at Queen's College, Marie told me about the exciting multidisciplinary mission of the Deep Carbon Observatory (DCO). I was immediately attracted by the vast transformative scope of this large-scale research programme with its focus on four clearly defined themes to be undertaken by four scientific communities working collegially. Through this framework the DCO had already completed six years of comprehensive exploration of deep carbon in Earth's crust, mantle and core. By this point in our conversation I was beginning to wonder how on Earth I could have anything to offer, given that my limited experience as a historian of science had been all about the pioneers who looked up in wonder at the mechanism of the heavens. So I was delighted to receive an invitation to participate as a historian of science at a DCO planning meeting soon to be held in a rural retreat at the University of Rhode Island in late 2015.

Next to a crackling campfire, Robert Hazen, through whose vision the concept of a large-scale enquiry on the multiple roles of carbon in Earth had emerged in 2007, passionately explained to me, "Simon, we need a proper history of deep carbon science." Such a history should identify 100 pioneers of deep carbon science; he would later give me a list of these, to which I would add another 50 or so. The narrative style should be one of telling engaging stories from history about how scientific enquiry is actually done: its contexts, circumstances, challenges and chance, as well as of the development of the scientific method. Could I identify sociological trends in the evolution of today's international interdisciplinary big science from the assortment of isolated individual scholars in the past? I have always liked unusual assignments that steer me along a new path of academic enquiry. This one could be fun. But more than that, I saw a door opening on a fabulous community of scholars deeply committed to research of the highest quality and wanting to share their discoveries with everyone immediately. Without reservation I agreed to give it a crack, which called for another beer while contemplating the embers of the campfire! I rose to the challenge, and this book is the result.

Historians work in the past – "a foreign country where they did things differently," as the saying goes. As a historian my task is to describe that foreign country and to account for how scholars and pioneering researchers made their discoveries under circumstances not at all like those that exist today. Mine is a deep history, by which I mean that I do begin at the beginning, insofar as that is possible. I decided that in order for us to a history of deep carbon science we need a narrative that outlines the long history of the discovery of the inner workings of our planet – the dynamics, the physics and the chemistry – and what we can learn about those things by measurements we can make at the surface, given that the interior is inaccessible. The slow accumulation of knowledge on the physical aspects of the interior set the scene for the twentieth century. In the geosciences, the tectonophysics revolution of the 1960s stands alongside the discovery of the structure of DNA in 1953 and the detection of the cosmic microwave background in 1965: all three were transformative for their fields. The stratigraphic column of my history tops out around the beginning of the third millennium. I have endeavoured to provide a solid platform from which future historians will survey the achievements of this century's pioneers of deep carbon science.

I   Why Carbon in Earth Matters

⊙ FOUNDATION OF THE DEEP CARBON
   OBSERVATORY

Carbon is the fourth most abundant element in the universe. It is outweighed by hydrogen, responsible for nine-tenths of the mass of ordinary matter in the cosmos, and by helium. Hydrogen and helium are remnants of the Big Bang: they are products of the first three minutes of our fireworks universe. Oxygen, the third most abundant element, and carbon are ashes from the explosive finale of the evolution of stars.

In 2009, Robert (Bob) Hazen and his colleagues of the Carnegie Institution of Washington promoted the following connection between carbon in the universe and human existence on Earth:

> Carbon plays an unparalleled role in human life. It is the element of life, providing the chemical backbone for all essential biomolecules. Carbon-based fuels supply most of society's energy, while small carbon-containing molecules in the atmosphere play a major role in our variable and uncertain climate. Yet in spite of carbon's importance scientists remain largely ignorant of the physical, chemical, and biological behavior of the carbon-bearing systems more than a few hundred meters beneath Earth's surface.[1]

Hazen et al. observed that we know neither how much deep carbon is stored in Earth's interior as a whole nor how deep carbon migrates along the pathways between the reservoirs. Furthermore, our ignorance of the deep microbial system – that by some estimates rivals the total surface biomass – is profound. In short, our knowledge of deep carbon is seriously incomplete. To address this knowledge deficit, in 2009 the Carnegie Institution of Washington launched a decade-long program of research and discovery: the Deep Carbon Observatory (DCO). Its mission was to lay the groundwork for a new scientific discipline devoted to element number six – carbon – and its place in our lives and world. The emergence of this new collaboration in the geoscience community has changed how science is conducted across time zones, cultures and disciplines to bring global thinking to bear on the role and properties of carbon inside Earth.

In 2007, a chance encounter between Bob Hazen and Jesse Ausubel, a faculty member of Rockefeller University and a project officer at the Alfred P. Sloan Foundation, led to the concept of the DCO. Hazen, an accomplished writer of exhilarating science books at the cutting edge of research, was on a promotional tour. Hazen gave an after-dinner talk at the Century Association, a club for "congenial companions in a society of authors and artists" on Manhattan's West 43rd Street. At this literary salon, Hazen spoke of his latest book, *Genesis: The Scientific Quest for Life's Origins*, in which he suggested that geophysical reactions might have played a critical role in getting life started. Ausubel's presence at this fundraising dinner was due to a last-minute cancellation by another participant. The presentation on the emergence of the first life on Earth made a deep impression on Ausubel. Hazen had developed a thinking style that envisaged life inevitably emerging as a consequence of chemistry, starting with water, organic molecules and a source of energy. His experiments in prebiotic chemistry showed the circumstances through which organic molecules could progress from structural simplicity to considerable complexity. This research was focused on how a prebiotic world rich in organic molecules could transition to the so-called RNA world of self-replicating genetic molecules. But above all, Hazen had emphasized the daunting gaps that existed in our knowledge of the origin of life and the special role of carbon. What Ausubel did next was to seek out Hazen's book and read it.

Three months later, Ausubel contacted Hazen about the possibility of the Alfred P. Sloan Foundation supporting an integrated science approach to the pursuit of life's origins. It would be a 10-year mission that drew on several branches of science – geology, biology, chemistry, physics and astronomy – in order to coordinate a multifaceted project to investigate early life on Earth and the role of the deep carbon cycle in its emergence. The first step was to convene an international workshop on the deep carbon cycle at the Carnegie Institution in May 2008. By this stage, Ausubel and Hazen were no longer focused solely on life's origins: they felt that, in order to further human understanding of Earth and our place here, they needed to place the element carbon center stage. In his opening address at the three-day workshop, Hazen set out what he wanted the 110 participants to achieve:

> A rare and important opportunity awaits us to define a new field. Collectively we need to assess what we don't know. We will succeed in this endeavor if we accomplish three things, which I now charge all of you with. First, we have to look beyond our individual interdisciplinary expertise and see the subject in an integrated context: geology, chemistry, biology, physics – they are all going to play central roles. Second, we have to identify the key questions we want to have answered to understand the deep carbon cycle. That's really what we're here to do. And finally, we have to imagine what it's going to take – what field observations, what key experiments, what new instruments, what theoretical advances are required to move this endeavor forward? I am tremendously excited to be here with you! I welcome you all! Let's get started!

All three tasks contributed to the publication three years later of *Carbon in Earth*, a monumental book that is the benchmark for our present understanding of Earth's carbon and a comprehensive review of what we already knew in 2009 and what we would have to learn in the DCO decade of discovery, 2009–2019.

This history of science book, *From Crust to Core*, complements *Carbon in Earth* by exploring four centuries of philosophical and scientific inquiry on the nature of Earth's interior, its cycles and mechanisms and the particular roles of deep carbon. My aim has been to present a layered story of remarkably rich discovery. The narrative thread encounters about 150 pioneers of deep geoscience, several dozen research institutions and universities and more than 20 ships and research vessels. Many of these pioneers started their inquiries from points of view that at first sight might seem far away from the history of deep carbon science. On the other hand, such personal journeys of discovery are central to the philosophy of science, showing how science is actually done and the importance of asking the right questions and of understanding what the data are actually telling us.

⊙ SPHERES BELOW AND HEAVENS ABOVE

Before commencing the historical narrative about the scientific discovery of Earth's deep interior, I shall introduce the architecture of Earth from crust to core as we understand it today. Readers already familiar with the concepts may wish to skip this section. The first point to make is that everything that is deeper than about 10 kilometres is inaccessible to direct view. Furthermore, the temperature rises surprising quickly. In the world's deepest gold mine, TauTona in South Africa, the temperature of the rock face is 60°C. At the bottom of the deepest borehole, 12 kilometres down, on the Kola Peninsula in Finland, the temperature reached 180°C, and which point further drilling became impossible.

Earth's interior is divided into layers with different chemical compositions and mechanical properties. To picture the internal structure of Earth, we can begin with simple models. It can be likened to a stone fruit such as an avocado: both have a solid core surrounded by a thick mantle, with a crinkly surface skin or crust. For geophysical purposes, however, the core–mantle–crust model is too crude. To improve on that, we must think of Earth as being made up of a number of layers, like an onion: we can peel them off one by one, starting with the crust, below which there are two layers for the upper mantle and the lower mantle, separated by the transition zone between depths of 410 and 660 kilometres. The transition zones (or boundaries) are where phase changes occur in minerals as we proceed

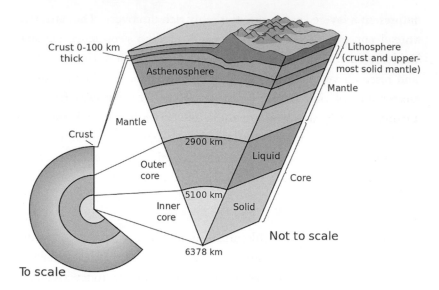

FIGURE I.I A conventional radial section of Earth's interior showing the major divisions used in geophysics.
Source: Adapted from the United States Geological Survey. Public domain

to greater pressures and higher temperatures. The boundary between the lower mantle and the outer core lies at a depth of 2900 kilometres. Below this is a liquid outer core, with a thickness of about 2300 kilometres, composed mainly of iron and nickel. The outer core is the seat of Earth's magnetic field, which is generated through a self-induced dynamo process. The transition to a solid inner core is located 5100 kilometres below the surface. We're going to examine these layers by working down from crust to core and then upwards from the surface to space.

Figure 1.1 shows the major divisions used in geophysics. The surface rocks are part of the crustal layer, which is rich in silica (silicon dioxide, SiO_2). Its average thickness is about 38 kilometres beneath the continents and around 8 kilometres beneath the oceans. The five commonest elements in Earth's crust are oxygen (47 percent), silicon (27 percent), aluminum (8.1 percent), iron (6.3 percent) and calcium (5 percent). Carbon (0.18 percent) is ranked tenth by its natural abundance in the crust. The thinner oceanic crust and the thicker continental crust are formed by entirely different processes and have different histories.

The crust and the uppermost mantle, considered as a mechanical entity, is known as the lithosphere (from the Greek *lithos*, meaning "rock"). It is the hard and rigid outer layer of Earth that has fractured into a dozen major plates (plus a handful of minor ones). Each tectonic plate is a layer of continental crust or oceanic crust supported by the viscous upper mantle. The oceanic lithosphere ranges in thickness from 50 to 150 kilometres, being at its thinnest at the mid-ocean ridges. The continental lithosphere is altogether a bulkier affair at a thickness of 40–280 kilometres or so, with a 30–50-kilometre veneer of crust. The boundary between the crust and the mantle is referred to as the Moho, a convenient contraction derived from the name of the pioneering Croatian seismologist Andrija Mohorovičić (1857–1936). In 1909, he first noticed a discontinuity in the behavior of seismic waves crossing the Moho.

The 100–200-kilometre-thick semi-fluid layer of hot plastic rock in the upper mantle is known as the asthenosphere (from the Greek *asthenēs*, meaning "weak"). Earth's layers vary in thickness, mechanical strength and chemical composition. Actually, there are two different concepts of layering in the outer part of Earth: the crust and the mantle have different compositions (geochemistry), whereas the lithosphere and asthenosphere have different mechanical strengths (geophysics). Under the influence of long-term stress, the lithosphere exhibits rigidity, but deforms elastically and through brittle failure, whereas the asthenosphere deforms like a highly viscous fluid. The multiple layers of the mantle arise because as we go from crust to core we encounter ever-increasing temperatures and pressures. Minerals in the mantle adjust their atomic structures and chemical compositions in reaction to different temperature and pressure regimes. Such phase changes are detectable because they alter the velocities at which earthquake waves travel through the interior. A transition zone between 410 and 660 kilometres marks the boundary of the lower mantle and the upper mantle. Much of our knowledge of the mineralogy and composition of the mantle has also come from experiments with diamond anvil cells and from microscopic examination of inclusions in diamonds.

To complete this introductory survey of our dynamic planet, we need to rise above the interior and consider four interconnected spheres: lithosphere, hydrosphere, biosphere and atmosphere. Together these make up the complete system in which life on Earth exists. The system is a physical and biological domain that is the subject of many great debates on environmental issues, climate change and the origin and evolution of life. The hydrosphere encompasses the water on, under or above the surface. So, it includes the water in the oceans and seas, the liquid and frozen groundwater, the water locked in glaciers, icebergs and ice caps and the moisture in the atmosphere. Three-quarters of Earth's surface is covered by oceanic saltwater – freshwater accounts for only 2.5 percent of Earth's surface, and just a tenth of that is readily available from lakes, reservoirs and rivers. The hydrosphere is an intricate closed system in which water and other volatiles are continuously driven around in a cycle powered by solar energy and Earth's gravity. This cycle moves water between the biosphere, atmosphere and lithosphere.

The atmosphere is the gaseous layer – commonly known as air – that surrounds Earth and is retained by gravity. By volume, dry air is 78 percent nitrogen, 21 percent oxygen, almost 1 percent argon and 0.04 percent carbon dioxide. Atmospheric scientists distinguish several layers in the atmosphere according to temperature and composition, as illustrated in Figure 1.2. The origin and evolution of the atmosphere is intimately connected to the interior dynamics of our planet. With the exception of the abundant oxygen released by photosynthesis, the atmospheric gases came from Earth's interior and were released through volcanic eruptions. Carbon dioxide is abundant in volcanic gases, which raises the question: Which emits more carbon dioxide – Earth's volcanoes or human activities? Terry Gerlach, a retired expert on volcanic emissions and formerly of the

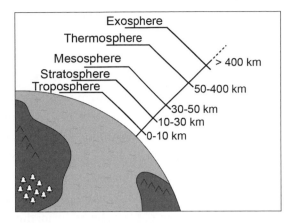

FIGURE 1.2 The principal divisions of Earth's atmosphere.
Source: Penny Wieser

United States Geological Survey, estimates that leakage from volcanoes currently amounts to 0.15–0.26 billion tonnes of carbon dioxide a year, whereas the anthropogenic contribution is more than 100 times greater than that.[2]

Just one more sphere remains to complete the series: the magnetosphere, a magnetic envelope surrounding Earth. The atmosphere and magnetic field protect us from most radiation hazards. These hazards include high-energy particles spawned by violent astrophysical events throughout our galaxy, as well as ejections of mass from the solar corona and the particles in the solar wind that breeze past at velocities of 400 kilometres per second. The geomagnetic field exists thanks to a dynamo mechanism: heat flow from Earth's inner core drives turbulent convection of fluid iron in the outer core, creating electric currents that produce a magnetic field. The mechanism is self-sustaining so long as there is sufficient energy to maintain convection.

⊙ LOOKING DOWN ON EARTH

To set the scene for this deep history, I shall begin in the period following World War II, when many physical scientists and engineers, exhausted by military research, returned to their laboratories and universities. What follows is one of several backstories leading up to the dramatic discovery of a dynamic Earth in the mid-1960s. In terms of the philosophy of science, I am including here a brief summary of the coincidental discovery of the fossil microwave radiation from the Big Bang event 13.7 billion years ago. During a few months in 1965, planetary and space science underwent enormous changes of human perspective – paradigm shifts, if you will. Twin revolutions of thinking upended our knowledge of the deep history of the universe and Earth and introduced a holistic systems approach to understanding how life, Earth and its environment, the universe and the natural world are interdependent and interlocked across multiple timescales.

In 1948, Fred Hoyle (1915–2001), an independently minded British astronomer and cosmologist, made this startling prediction in a radio broadcast:

> Once a photograph of the Earth, taken from outside, is available ... a new idea as powerful as any in history will be let loose.[3]

He later looked back with great pleasure on his forecast. On January 6, 1970, in his banquet speech to the First Lunar Science Conference in Houston, Texas, he proclaimed:

> Well, we now have such a photograph. I've been wondering how this old prediction stands up. Has any new idea in fact been let loose? It certainly has ... everybody has become seriously concerned to protect the natural environment. Something unique has happened to create an awareness of our planet as a unique and precious place.[4]

The first such photograph was a fuzzy 1966 monochrome image taken by the spacecraft *Lunar Orbiter 1*. The environmental game changer came two years later with the December 1968 *Earthrise* photograph taken during the *Apollo 8* mission: Figure 1.3 shows our entire world as a small, blue, finite globe, with our nearest celestial neighbor, the Moon, a desolate presence in the foreground. Overnight

FIGURE 1.3 *Apollo 8* monochrome image of Earth rising above the limb of the Moon, the first view of Earth from another planet, taken by astronaut William Anders. *Source:* NASA Earth Observatory, image remastered 2013. Public domain

the popular consciousness of people worldwide gasped at the fragility of our existence in the immensity of the cosmos. The perception of our isolation on a small world in orbit around an ordinary star profoundly altered human perceptions of our place in the universe. Humanity grasped the opportunity to find out how our planet functions as a whole system. Environmental movements compelled humans as a species to understand how we have reshaped the world. A new interdisciplinary field – Earth system science – flourished, nurtured by populist support. The 1970s saw the emergence of a new political ideology in many countries. As such, the nature photographer Galen Rowell promoted the image as "the most influential environmental photograph ever taken." From 1990, Carl Sagan (1934–1996) famously championed the environmental context of extraterrestrial photographs of terra firma with *Pale Blue Dot*, a photograph of Earth taken by *Voyager 1* from a distance of about 6 billion kilometres. So, when we first looked down on Earth from space, Hoyle's powerful new ideas were unleashed in the half-century from the mid-1960s to the present day.

Historically, 1963–1968 was a tumultuous period of revolutionary discovery in the Earth and space sciences. In 1963, Frederick Vine, a doctoral student of geophysics at Cambridge, together with his supervisor Drummond Matthews, put forward the hypothesis that sea-floor spreading at mid-ocean ridges could explain magnetic anomalies in the rocks of the ocean floor. The magnetism of the rocks had recorded the story of continents drifting apart, propelled by convective thermal energy from the deep. This seemed like madness at the time. Today we have a good understanding of the evolution of Earth, commencing with its hot formation about 4.6 billion years ago through accretion in the solar nebula. In early 1965, the progress of cosmology soared dramatically following the discovery of the cosmic microwave background: this fossil radiation is very dilute thermal energy from the hot Big Bang during the explosive origin of our universe 13.7 billion years ago. Hoyle praised the remarkable photograph from *Apollo 8* five years after the astronomers' discovery of a vast and dynamic universe above our heads and the geophysicists' discovery of a dynamic inner world beneath our feet.

☉ THE INVENTION OF EARTH SYSTEM SCIENCE

The properties of our expanding universe are now so well established that a concordance cosmology emerged at the beginning of the twenty-first century. We have discovered with impressive precision the basic parameters of our universe, such as its age, its expansion rate and the properties of its mass and energy. And we are aware of a great deal about its contents: galaxies and clusters of galaxies, dark matter and dark energy, fossil radiation and gravitational waves, supermassive black holes, stars and interstellar gas, planets and their moons. We now have insights into how the various cosmic components evolve through time. These developments in astronomy became significant in the Earth sciences because it became possible to describe the history of matter, from the origin of the universe to the formation of the solar system. This theme is explored in Chapter 2.

The 50 years from Einstein's general theory of relativity in 1915 to the detection of the cosmic microwave background witnessed phenomenal progress in the philosophy of science. Cosmologists forged ahead, advancing from dreamy speculations about our universe being static or dynamic and arriving with total certainty at the conclusion that it is expanding. Observational astronomers broke out of the narrow bandwidth of the visible spectrum, swinging open new windows to the radio and X-ray universe. In the second half of the twentieth century, astronomers self-organized into large collaborations and networks, supported by public funding of space probes to the planets and great observatories in Earth orbit, such as the Hubble Space Telescope. By the dawn of the new century, the publication of data from large surveys and from space telescopes of all kinds had become open access, freely available to everyone.

While the astronomers probed ever further into outer space, the Earth scientists dived into our inner world. In a paradigm shift that mirrored what had happened in cosmology, new concepts developed rapidly, starting with plate tectonics. Yes, continents are adrift, with Europe and America distancing themselves by about 3 centimetres a year and India steadily crashing into Eurasia. A global recycling and reassembly scheme is in place. Steady subduction of the sea floor occurs at the continental margins, and the eruption of lava at the mid-ocean ridges marks the parting of the ways. Geoscience experienced a marked change of emphasis: geologists laid aside their hammers and instead embarked on ocean-going research ships to crack open the secrets of the hoary deep. Seismology had its shake-up, too, when exploration geophysics morphed from searching for hydrocarbon deposits of commercial interest to unlocking the details of Earth's deep internal structure and dynamics.

The minerals in rocks tell us stories about their history: how, when and why they formed and to what extent they have been transported and transformed over deep time. To study how rocks and minerals behave deep down, geoscientists are able to replicate the high pressures and temperatures of Earth's interior in their laboratories. Remarkably, this can be accomplished at the lab bench with a thumbscrew and two diamonds. Here's how: in a diamond anvil cell, two brilliant-cut diamonds with perfectly flat faces are jammed head to head with a minute sample of rock between them. Diamond is such a hard form of carbon that it can withstand pressures up to hundreds of thousands of times (or more) greater than the atmospheric pressure at the surface.

From 1968, mineralogy expanded rapidly by combining the diamond anvil cell with the microscope in order to use a wide range of imaging and spectroscopic techniques to probe how the physical and chemical properties of minerals can change during geological processes. The study of minerals is a fundamental aspect of geophysical inquiry. Minerals existed before there were any forms of life on Earth. Indeed, minerals played an important role in the origin and evolution of life, interacting with biological systems in ways that we've only recently started to recognize. In the dynamic interior of our planet, the behavior of minerals as they churn their way through high-pressure and high-temperature regimes is the key to understanding the physical conditions in the deep Earth and how they have changed over geological time.

⊙ ENVIRONMENTALISTS SHAKE UP HUMANITY

When we first looked back at Earth from space, James Lovelock, a British life scientist, was working for NASA. He collaborated on planetary exploration research with colleagues at the Jet Propulsion Laboratory (JPL), Pasadena, California. This was an exciting time at JPL: the Viking missions to Mars were to land two spacecraft safely on the red planet. Lovelock's job was to develop sensitive instruments for the analysis of the atmosphere and surface of Mars. This task was motivated in part to determine whether Mars supported life. In 1965, however, Lovelock took a hard look at the Martian atmosphere with its overwhelming abundance of carbon dioxide. Could that atmosphere really support life? Probably not, he concluded, noting its sharp contrast to the chemically dynamic mixture of nitrogen, oxygen and carbon dioxide in Earth's biosphere that brimmed with life.

Through his research for NASA concerning life on Mars, Lovelock formulated an audacious hypothesis. After discussions with Carl Sagan, Lovelock announced that the composition of Earth's atmosphere is largely a consequence – a by-product, as it were – of life on Earth. This Gaia hypothesis (or theory, as it is now known) is named after the Greek Earth goddess. Its claim is that life and its non-living environment is a self-regulating system that ensures Earth's climate and the composition of its atmosphere are always suitable for life to continue. It's an example of what cosmologists would later refer to, by turns, as fine-tuning or the anthropic principle; that is to say, the world is as it is because we are here to observe it. Lovelock developed this worldview in the 1960s and 1970s through cooperation with the American evolutionary theorist Lynn Margulis (1938–2011). The Gaia theory is the first scientific statement of Earth as a symbiotic system that is more than the sum of its parts. Lovelock was familiar with systems theory and cybernetics. As a result, he had a high level of familiarity with positive and negative feedback. He felt that a system as complex as Earth must have a multitude of such loops. From this point, the identification of feedback loops in Earth's natural systems became a key driver of Earth system science.

The Gaia theory did not gain deep and wide acceptance in the scientific community, probably because associating it with Greek mythology and philosophy was a serious category error, but it did capture public interest. It boosted the environmental movement that really took off in the USA largely thanks to the advocacy of scientists such as marine biologist Rachel Carson (1907–1964). Her book *Silent Spring* (1962) dealt with environmental problems that she believed were caused by synthetic pesticides such as dichlorodiphenyltrichloroethane (DDT). The overriding theme of *Silent Spring* is the powerful – and often negative – impact humans have on the natural world. Hers was the first voice to spread concern that the human use of chemicals could interfere with the biosphere on a global scale.

Other areas of concern about pollution of the Earth system rapidly followed. Lovelock's laboratory research led to him becoming the first to detect the global build-up of chlorofluorocarbons (CFCs) in the stratosphere. CFCs are organic compounds containing carbon, chlorine and fluorine. They were widely used as refrigerants, aerosol propellants and solvents. In 1974, two atmospheric chemists, Mario Molina and Sherwood Rowland (1927–2012), found that CFCs were depleting the protective ozone layer in the stratosphere that absorbs most of the Sun's ultraviolet radiation: long-lived chlorine ions react with ozone to form long chains of molecules composed of chlorine and oxygen. They received the Nobel Prize in Chemistry in 1995 for this discovery. By 1987, the Montreal Protocol set strict limits for CFC emissions. Even so, it will take until 2075 for the polar ozone layer to recover completely. Our use before the 1980s of refrigerants based on carbon and halogens has changed the chemical make-up of the stratosphere to such an extent that full recovery of the Earth system will take nearly a century. In the 1980s, rising atmospheric pollution by another gas, carbon dioxide, also led to calls for global action to mitigate the great global warming crisis.

⊙ GLOBAL WARMING AND DEEP CARBON

By the middle of the nineteenth century, a few scientists were aware that climate change must have occurred, simply to explain Ice Ages. The outstanding Irish physicist John Tyndall (1820–1893) was a keen alpinist: he was among the first to ascend the Weisshorn (4506 m) and was an early climber of the Matterhorn (4478 m). His familiarity with glaciers, on which he wrote a major monograph, convinced him that tens of thousands of years ago northern Europe must have been entirely buried under colossal amounts of ice, and therefore the climate must be subject to long-term change. He was aware that Joseph Fourier (1768–1830) in Paris had suggested in 1824 that Earth's atmosphere kept the surface temperature warm by acting as a blanket blocking transfer of the Sun's radiant heat to the vacuum of space.[5] As a keen experimentalist at London's Royal Institution, Tyndall decided to find out if there were any gases in the atmosphere that could trap infrared radiation. And yes, there were three: he identified water vapor as the most important, but he also found that carbon dioxide and methane as trace gases are amazingly effective at altering the balance of heat radiation through the entire atmosphere.

Towards the end of the nineteenth century, Swedish physical chemist Svante Arrhenius (1859–1927) became fascinated by the riddle of the historical cause of Ice Ages, which had become a hotly contested topic at the Stockholm Physical Society. He was the first scientist to model the effect of atmospheric carbon dioxide on global warming. His model suggested that doubling the amount of carbon dioxide would raise the temperature in Europe by 5–6°C.[6] His colleague Arvid-Gustave Högbom (1857–1940), also at the University of Stockholm at the time, had a strong interest in the global carbon cycle. He estimated that the flux of carbon dioxide from the industrial burning of coal (a form of deep carbon) was comparable to that from natural geochemical sources such as volcanoes. In a popular book published in 1908, Arrhenius suggested that a doubling of carbon dioxide would take many centuries. In any case, the concept of warming was naturally attractive to Scandinavian scientists, and the German physicist Walther Nernst (1864–1941) even suggested setting fire to unused coal seams to boost the global temperature. Tyndall, Fourier, Arrhenius and Högbom all contributed to the study of the effect of carbon dioxide atmospheric thermodynamics as an important aspect of Earth system science. However, the realization that anthropogenic carbon dioxide emissions would have massive implications for the future habitability of the planet did not come in their lifetimes.

In 1958, Charles David Keeling (1928–2005), an early career scientist at the Scripps Institute of Oceanography in San Diego, invented a detector for sniffing the air and measuring the concentration of carbon dioxide. On November 22, 1958, he set in train the daily monitoring of carbon dioxide outgassing on the north flank of Mauna Loa, the largest volcano in the world, topping out at 3397 m and at a great distance from major sources of pollution. The Mauna Loa Observatory has continued hourly measurements of atmospheric quality ever since. Keeling's initial results gave a concentration of 310 parts per million (ppm). When he died this number had risen to 380 ppm, and at the time of writing (2018) it has soared to 420 ppm. On April 25, 1969, Keeling took his findings of the rise in carbon dioxide to a symposium at the American Philosophical Society on the long-term medical implications of atmospheric pollution.[7] His paper was preceded by one on lead pollution and followed by one on air pollution and plant life. He was writing at a time of severe smog pollution in Los Angeles and San Diego, which had adversely affected him and his family. Here's the final sentence from his paper:

> If the human race survives into the twenty-first century with the vast population increase that now seems inevitable, the people living then ... may face the threat of climate change brought about by an uncontrolled increase in atmospheric CO_2 from fossil fuels.

Historically, that's one of the first hard-hitting statements about the unexpected global consequences of the accelerated use of deep carbon as a source of energy for industrial and domestic purposes.

⊙ JULES VERNE IMAGINES EARTH'S DYNAMIC INTERIOR

In concluding this chapter, I would like to comment on Jules Verne's second epic adventure novel, the scientific romance *Voyage au centre de la Terre*, published by Pierre-Jules Hetzel in 1864. It is an impressive example of the use of science fiction to promote the latest discoveries and their possible applications. In Verne's day, divergent theories about the nature of Earth had scholars disputing whether Earth's interior was liquid or solid. Estimates of the age of Earth were subject to stupendous variations. Darwin's theory of biological evolution over immense periods of time, together with the geological evidence for the development of Earth over countless millions of years, challenged the literal descriptions of creation in biblical texts. In the mid-nineteenth century, a curious public and enthusiasts became excited by public lectures and popular science expositions on Earth's distant history. Dinosaur fossils began to be displayed in museums. Verne, and other writers, tapped into this interest in the past by crafting engaging narratives in which they shared controversial ideas about extinction, evolution and a world once dominated by reptiles.

Throughout *Voyage au centre de la Terre*, Verne's familiarity with the geoscientists of his day and their theories shines through impressively, thanks to his scrupulous research. Verne's inspiration came from Scottish geologist Charles Lyell (1797–1875), who had investigated volcanoes and earthquakes. Verne learned about volcanic phenomena from Charles Sainte-Claire Deville (1814–76), a geologist at the Collège de France, who had made a geological study of Tenerife (1848) and had witnessed eruptions of Stromboli, the setting for the conclusion of the novel. Verne cites Georges Cuvier (1769–1832) five times in connection with his research on fossils that established the extinction of species as a fact, as well as Cuvier's opposition to evolution.

REFERENCES

1. Hazen, R. M., Hemley, R. J. and Mangum, A. J. Carbon in Earth's interior: storage, cycling, and life. *Earth and Space Science News* **93**, 17–18 (2012).
2. Gerlach, T. Volcanic versus anthropogenic carbon dioxide. *Earth and Space Science News* **92**, 201–208 (2011).
3. Hoyle, F. *The Nature of the Universe* (Basil Blackwell, 1950).
4. Hoyle, F. Banquet speech at the first Lunar and Planetary Science Conference. St. John's College, Cambridge Special Collections, Papers of Sir Fred Hoyle (1915–2001), unpublished typescript, container 5 (1970).
5. Fourier, J. Mémoire sur les températures du globe terrestre et des espaces planétaires. *Mémoires de l'académie royale des sciences* **7**, 569–604 (1827).
6. Arrhenius, S. On the influence of carbonic acid in the air upon the temperature of the ground. *Philosophical Magazine and Journal of Science* **41**, 237–276 (1896).
7. Keeling, C. D. Is carbon dioxide from fossil fuel changing man's environment? *Proceedings of the American Philosophical Society* **114**, 10–17 (1970).

2 The Origin of Deep Carbon in Deep Space

⊙ CARBON IS UNIVERSAL

Carbon is the fourth most abundant element in the universe, and it is one of the most important elements on our planet. In this chapter, we introduce the story of Earth's carbon all the way from its synthesis in the first generation of stars in our universe, to its incorporation in the solar nebula, where the Sun and planets formed nearly five billion years ago. Carbon's journey from deep space to deep Earth took almost nine billion years. It is the basis of all life on Earth, where it serves as the structural backbone of molecules large enough to carry biological information. One of carbon's most important features is that it readily forms chemical bonds with many other atoms. This property is the driver behind the biochemical reactions needed for metabolism and propagation. The history of life on Earth is therefore inextricably linked with the history of these elements.

The compound noun *deep carbon* refers to terrestrial carbon that is not in the atmosphere, the oceans or on the surface. In simple terms the meaning is: carbon and carbon systems of any kind that are a few hundred metres beneath our feet. Our scientific understanding of deep carbon is seriously impaired. To address this knowledge deficit, in 2009 the Carnegie Institution of Washington launched the Deep Carbon Observatory as a decade-long global research program, organized around four themes of inquiry, each devoted to a particular aspect of carbon in Earth's interior. These areas of inquiry can be neatly set out as a series of questions requiring answers. Where is carbon sequestered (reservoirs) and what is the rate of change over time of the deep carbon inventory (fluxes)? How do carbon-bearing minerals behave when subjected to the extreme physics and chemistry of the interior? What can we learn about deep sources of carbon-based fuels (deep energy)? And what can we discover about the deep biomass in the ocean crust (deep life)? This book tells the story of the more than four centuries of discovery and debate leading up to the Deep Carbon Observatory and to the invention of a new interdisciplinary field – deep carbon science. To do that, I'm beginning at the beginning, as historians should, by answering the question "Why is there carbon?"

⊙ IT STARTED WITH A BIG BANG!

To explain the origin of the chemical elements, I'll start with cosmology, the field of study that considers the universe on the largest scale (its entirety) and the longest time (from its origin to the present day). For more than 2000 years, philosophers had pondered the meaning of the ever-changing night sky. That type of intellectual activity is known as metaphysics. The sages of ancient Greece and the medieval theologians tried to answer two general questions: What is there and what is it like? They were not doing science, and they did not seek underlying causes: for them, merely accounting for the *appearance* or *essence* of the heavens was good enough.

This state of affairs endured in cosmology until the middle of the fifteenth century. That's when Nicolaus Copernicus (1473–1543) daringly proposed a hypothesis with the Sun at the center of the universe. In 1623, the Italian polymath Galileo Galilei (1564–1642) set in motion the transition to modern science. He ordained that natural philosophy (physics) should employ mathematical laws rather than the teachings of the ancients. In his middle years, Isaac Newton (1643–1727) proposed the laws of motion and gravitation to account for the motion of celestial bodies. Newton's laws held sway until November 1915, when Albert Einstein (1878–1955) introduced an entirely new theory of gravity (general relativity) that could be applied to the universe as a whole.

Einstein wrapped up an extraordinary revolution in physics. Within another 50 years, astronomers had discovered the vast size of the universe and established that it is expanding. In the late 1940s, the hypothesis that the origin of the universe was in an event known as the Big Bang gathered momentum. Finally, in 1965, radio astronomers serendipitously discovered the fossil radiation from that primordial event. Astronomers nevertheless needed another four decades to uncover incontrovertible evidence that 13.7 billion years ago the universe suddenly appeared out of nowhere!

The Big Bang produced all of the matter and radiation that the universe contains today. Within one second its temperature fell from ~10^{28} K to 10^{10} K. Neutrons and protons then froze out of the plasma. In the next 100 seconds the temperature plummeted to 10^9 K, cold enough for a neutron and a proton to combine to form a deuteron. Almost all of those deuterons then paired up to form helium nuclei. Very soon there were no neutrons because the half-life of a neutron is 611 seconds: for stability the neutron needs to be confined to an atomic nucleus, where an immensely strong force binds together protons and neutrons. By mass, the quantity of light elements assembled in the Big Bang amounted to about 75 percent ^{1}H (hydrogen), about 25 percent ^{4}He (helium), about 0.01 percent ^{2}H (deuterium) and ^{3}He and a trace amount of ^{7}Li (lithium). At this stage, the most common nuclei were ^{1}H and ^{4}He, and indeed, hydrogen and helium have remained the commonest elements in the universe ever since.

The Big Bang nucleosynthesis frenzy fizzled out after the first 1000 seconds. That's because the universe was not hot enough to fire up nuclear fusion reactions. Even though there was an abundance of electrons, the first atoms did not gel until the age of the universe reached 380,000 years. By then, the universe was cold enough for electrons to combine with protons to form hydrogen and helium atoms. This process of uniting electrons and nuclei swept the universe clean of the free electrons: it became transparent to radiation. Photons from this era have survived as cosmic microwave background radiation, which was first announced in 1965.

☉ STELLAR ORIGIN OF THE CHEMICAL ELEMENTS

When did the first stars and galaxies form? That question was first posed in ancient times. Today cosmology is a mature science based on exquisite observations that really do take us on a time travel ride back to the earliest galaxies. We can see them, so there's no need to speculate. The first billion years were a crucial epoch in the history of the universe, because that is when the first stars and galaxies formed. Our understanding of those galaxies has improved greatly thanks to the final upgrade of the Hubble Space Telescope. On May 14, 2009, two astronauts aboard the *Atlantis* Space Shuttle made a spacewalk lasting seven hours to install a very sensitive infrared camera and a replacement computer to handle data transmission. This upgrade pushed the detection limit for galaxies even deeper, and even further back in time. The big boost to performance enabled an international team of 18 astronomers to announce in March 2016 that the Hubble Space Telescope had imaged a large galaxy (Figure 2.1) of a billion solar masses that already existed less than 500 million years after the Big Bang. Galaxy assembly and star formation began no later than 200 million years after the Big Bang.[1]

FIGURE 2.1 Inset: the most distant galaxy in the universe, GN-z11, seen as it was 13.4 billion years in the past, just 400 million years after the Big Bang. This galaxy is ablaze with massive young blue stars, transforming hydrogen and helium into heavier elements..
Source: NASA, ESA and Space Telescope Science Institute. Public domain

An English chemist, William Prout (1785–1850), was the first scientist to pose the question, "What is the origin of the chemical elements?" In 1815, he published his hypothesis that the relative atomic mass of every element is an exact multiple of the mass of the hydrogen atom.[2] The following year, he suggested that all elements were formed from hydrogen by some process of condensation: "[W]e may also consider the [primordial matter] of the ancients to be realized in hydrogen."[3] However, by the mid-1830s, more precise measurements of atomic weights falsified that hypothesis: the atomic weight of chlorine, which is 35.45 times that of hydrogen, could not be explained. Prout's hypothesis faded into obscurity until the discovery of isotopes in 1913 bore out his idea.

The great astrophysicist Arthur Eddington (1882–1944) suggested in his popular book *Stars and Atoms* that the Sun is powered by nuclear energy. He outlined how four hydrogen nuclei (four protons) could fuse to create one helium nucleus, releasing energy through the conversion of matter.[4] In 1939, Hans Bethe (1906–2005) of Cornell University published a scheme for a catalytic cycle of nuclear reactions that could power the stars. Each turn of the cycle facilitated the interactions of four protons to produce one helium nucleus. The isotopes of carbon, nitrogen and oxygen were choreographed during the cycle to provide exit channels for the two positrons, the two neutrinos and the stupendous amount of nuclear energy that was superfluous to the stability of the helium nucleus. For this tremendous breakthrough in the synthesis of elements in stars, Bethe won the Nobel Prize in Physics in 1967. The abstract of his seminal paper states:

> It is shown that the most important source of energy in ordinary stars is the reactions of carbon and nitrogen with protons. . . . [C]arbon and nitrogen merely serve as catalysts for the combination of four protons (and two electrons) into an α-particle [helium nucleus].

This paper is highly significant historically: in it, Bethe established that carbon plays an essential role in the astrophysical

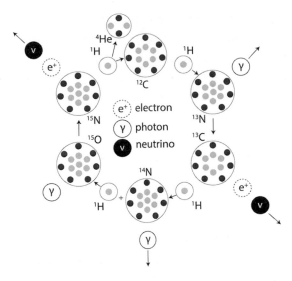

FIGURE 2.2 The carbon–nitrogen–oxygen (CNO) cycle is one of the two sets of nuclear fusion reactions occurring in stellar interiors that transform hydrogen into helium. It is the dominant reaction in stars of more than 1.3 solar masses. It is catalytic, with carbon, nitrogen and oxygen recycling in an endless loop.
Source: Penny Wieser

conversion of mass into energy. Figure 2.2 illustrates the CNO cycle: carbon (together with nitrogen and oxygen) is one of the three elements that turn Einstein's $E = mc^2$ equality into the existential transformation $E \rightarrow mc^2$ in stars. *Et voila!* May the stars shine brightly!

A few years later, early-career nuclear astrophysicist Fred Hoyle contributed another great discovery to the story of carbon in the universe. During 1945–1950, many physicists who had been engaged in the development of the atomic bomb productively turned their attention to the origin of the elements. Atomic weapons research had produced a mass of data on nuclear fusion reactions, in which two or more atomic nuclei come close enough to transmute into one or more nuclei and subatomic particles. We have just noted that in the Big Bang hydrogen nuclei fused to form helium. Nuclear physicists investigated how fusion reactions in the cores of stars might be able to build up the chemical elements in chain reactions. Of course, at the

FIGURE 2.3 Victor
Goldschmidt. Mineral and
Geological Museum,
Oslo, 1920s.

outset they understood that acceptable models of these processes
must replicate the relative proportions of the elements as are observed
in the cosmos.

Work on the chemical abundances of elements in the cosmos
relative to their atomic weights had already been compiled in the
1930s by the geochemist Victor Goldschmidt (1888–1947) and his
colleagues in Göttingen and Oslo. Goldschmidt (Figure 2.3) laid the
foundations of modern geochemistry, becoming highly praised by
nuclear astrophysicists for undertaking "a most careful and painstak-
ing analysis of all the existing geochemical and astrochemical data."[5]
Without Goldschmidt's meticulous work, they could not possibly
have cracked the problem of the nuclear origin of the elements.
Goldschmidt's plot of the abundances of the chemical elements
(Figure 2.4) can be considered as a Rosetta Stone for nuclear
astrophysics. They showed that roughly the same proportions of the
commoner elements such as carbon, oxygen, nitrogen, silicon, magnesium, iron and so on had had a common genesis. The burning
questions for the nuclear astrophysicists were: Where and how had
nature created the elements, and their isotopes, in the proportions we
see today in Earth's crust and the cosmos? Why was carbon so
common and gold so rare? Why are lithium and beryllium extremely
rare, given that they are in positions three and four in the periodic
table, whereas abundant element six, carbon, is just two steps away?

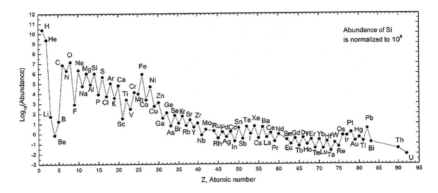

FIGURE 2.4 Natural abundances (log scale) of chemical elements (atomic number) in the solar system.

⊙ ORIGIN OF CARBON IN THE UNIVERSE

In the late 1940s, the starting point for researchers of cosmic cookery was to discover if the elements could have been formed in the Big Bang along with hydrogen and helium. However, they ran into an impassable bottleneck: in nature there are no stable isotopes with atomic mass 5 or 8. This is a crucial point: if a helium nucleus is battered by a neutron or two helium nuclei smash together, the resulting nucleus of mass 5 or mass 8 does not survive for long enough to participate in a further nuclear reaction. So how could heavier elements be built from lighter elements?

In 1946, Hoyle investigated a chain of nuclear reactions that he showed might explain element synthesis in the interiors of red giant stars, where the central temperature reaches 4 billion degrees. At that temperature, the nuclei have sufficient energy to fuse and then hold together. His aim was to discover the pathway for synthesizing elements up to iron. He ignored the bottleneck by starting with ^{12}C (carbon-12) as a given, to which he added ^{4}He (helium nucleus) for the synthesis of ^{16}O (oxygen-16). Add another helium nucleus to make ^{20}Ne (neon-20) and so forth up to ^{40}A (argon-40). To get traction for this scheme, though, Hoyle had to assume that carbon was made in the interiors of stars by a process unknown at that time. Clearly, it was not made in the Big Bang, but it was abundant in stars. He left this origin puzzle as unfinished business while he continued to work on stellar nucleosynthesis. At the same time, he built up a personal network of contacts in Pasadena: astronomers such as Edwin Hubble (1889–1953) at the Mount Wilson Observatory and the nuclear physicists at Caltech. Hoyle's academic style always involved a network approach, enabling him to learn from outstanding intellectuals with a passion for independent thought and skepticism for theories that were "generally accepted."

Hoyle spent a productive sabbatical year at Caltech in 1953, where the nuclear astrophysicist William (Willy) Fowler (1911–1993) was busily developing theories on how light elements were built up in stars. Fowler's group at the W. K. Kellogg Radiation Laboratory examined the nuclear energy levels of isotopes of carbon, nitrogen and oxygen. They looked for any pattern of regularity in the energy levels of light nuclei, seeking signatures that might reveal the inner structure of the nucleus. When Hoyle arrived at the Kellogg Laboratory, on Fowler's prior invitation, the experimentalists were focused on the nuclear properties of ^{13}C (carbon-13). That research was principally directed to the US atomic weapons program. An unintended offshoot was discovering the reaction for the synthesis of carbon in stars. Hoyle repaid his host with one postgraduate lecture a week on experimental cosmology, during which he expounded his thoughts on nucleosynthesis. Ironically, his background in experimental nuclear physics was rather weak compared to that of his Caltech audience. Whenever a sceptic interjected with, "That doesn't happen in stars," Hoyle dealt with his opponent on the fly with the invocation, "Let's stop here for today and we'll continue next time." By the following week he had overcome whatever the obstacle was, more or less. According to Ward Whaling, an assistant professor who was present at Hoyle's lectures, "[Hoyle] was sort of making this stuff up as he went ... it was quite exciting to see."[6] Great excitement came when Hoyle outlined the problem of getting from helium to carbon in one fast reaction and he experienced a sudden brainwave.

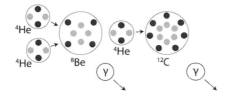

FIGURE 2.5 The triple-alpha reaction. Shown on the left: two alpha-particles (^{4}He) form ^{8}Be, which immediately absorbs a third alpha particle to form ^{12}C in an excited state, as is shown on the right. Initially, the Caltech nuclear physicists were highly skeptical of this proposal..
Source: Penny Wieser

In simple terms, it was already known that to build carbon from helium in *one* reaction required the near-simultaneous collision of *three* helium nuclei. That's because the helium nucleus has two protons and two neutrons, whereas carbon has six of each. A simultaneous hooking up of three helium nuclei would not work. Hoyle's delicate interpretation of the process was a two-stage affair. In the first stage, two helium nuclei collide to make one ^{8}Be (beryllium-8), which momentarily quivers in an excited energy state, but the shuddering nucleus does *not* decay instantly. It pauses for 10^{-16} seconds, just long enough to grasp a passing helium nucleus. That was Hoyle's hypothesis. But in reality, there's the problem of parking the nuclear energy released by the fusion of ^{8}Be and ^{4}He: it has the potential to shatter the new carbon nucleus.

In a flash of inspiration (which lasted far longer than 10^{-16} seconds!), Hoyle speculated that *if* the ^{12}C nucleus has an enhanced energy state, then the surplus energy could be stored at that level. He worked out how far above the ground state the enhanced energy level would need to be in order to survive the release of the fusion energy. Next, he needed to know if that energy level existed in the ^{12}C nucleus. And he was in exactly the right place to get the answer: the Kellogg Laboratory, where Fowler's group was cracking open the nuclear secrets of ^{13}C. Hoyle could scarcely contain his excitement.

The brash young Yorkshireman crashed into Fowler's office without so much as a "by your leave." At that stage, Fowler still regarded Hoyle as an eccentric cosmologist with pushy mannerisms who spoke with a strange Yorkshire accent. Fowler would not interrupt the scheduling of his new accelerator to accommodate a brusque Brit whose brain was brimming with an obsession over the energy states of ^{12}C. Hoyle persisted with his enthusiastic insistence to the point where Fowler convened a council of war in his cramped office. The consensus of Fowler's team was that *if* Hoyle's radical argument were correct – which seemed highly improbable – then the consequences for the theory of element synthesis in stars would be immense. "Let's do it!" they cried. Their experiment took three months. Hoyle could not hang about that long. He had to get back to basics in Cambridge, where a demanding undergraduate teaching schedule required his attention.

To their great surprise, the Caltech nuclear scientists did find the enhanced energy level in ^{12}C with precisely the value Hoyle had predicted! They put his name first on the paper announcing the discovery. Twenty years later, Fowler remembered:

> We then took Hoyle very seriously, because of this triumph from our standpoint of predicting the existence of a nuclear state from astrophysical arguments. I at least took him very seriously, and we did a lot more work on his state of carbon-12. We had to show, for example, that the resonance not only existed, but that it could be formed from three alpha particles.[7]

From the perspective of history, we can see that "this triumph" spurred Hoyle to revisit his paper of 1946, now that the origin of cosmic carbon was fully understood. The reaction (Figure 2.5) in which one carbon nucleus is assembled from three helium nuclei is known as the triple-alpha process. The sweeping title of Hoyle's 1954 paper in *The Astrophysical Journal* is "The Synthesis of Elements from Carbon to Nickel." Clearly, carbon was the basement rock on which the existence of all heavier elements rested.

In the same year, two young astrophysicists with doctorates from University College London, Geoffrey (1925–2010) and Margaret

Burbidge (1919–2020), arrived in Pasadena to work with Fowler. Margaret had a deep interest in what could be learned about cosmic chemical abundance by observing the spectra of stars. Her husband Geoffrey made the observations; Margaret was barred on sexist grounds from such "night work" at the Mount Wilson Observatory, where the all-male dormitory was known as the monastery. Enough said. Happily, downtown at Caltech, everything was now in place for one of the greatest collaborations in twentieth-century physics and chemistry. Working in a windowless visitors' office at the Kellogg Laboratory, around the corner from Fowler's cubbyhole, the three Brits churned away on hand-cranked calculators and chalked their ideas on a blackboard, trying to figure out the origin of all of the chemical elements. Their fact-finding included the public release of classified data on low-energy nuclear reactions and nuclear data from the second hydrogen bomb explosion, which had vaporized Enewetak, an island in the Bikini archipelago, on March 1, 1954. The four pioneers of element synthesis published their conclusions in 1957, running to 108 pages, as an encyclopedic paper in *Reviews of Modern Physics*.[8] The paper explained the origin of almost every isotope in the periodic table of the elements, including the rare earths that are so important in geochemistry.

I have often proposed that this paper, fondly referred to as B[2]FH by astrophysicists, is the most influential paper in astrophysics published between 1945 and the introduction of fast digital computers in the 1960s.[9] The paper marked Hoyle's greatest moment as a creative research scientist. Fowler received a half-share of the Nobel Prize in Physics in 1983 "for his theoretical and experimental studies of the nuclear reactions of importance in the formation of the chemical elements in the universe." Inside a star, the onset of element building begins in the core with the formation of carbon from three alpha-particles. That process marks the beginning of the history of carbon in the universe. And now we will take an amazing journey to follow the transfer of carbon from the cores of stars to deep inside Earth.

☉ EXPLODING STARS SEED SPACE WITH CARBON

In the early history of our galaxy, intermediate and massive stars, with lifetimes ranging from about 10 billion years down to a few millions of years, were the source of elements and their stable isotopes for everything heavier than helium. In the later life of such stars, hydrogen burning in the core ceases because it has become exhausted, and helium burning commences in earnest. That's when carbon-12 is made via Hoyle's triple-alpha process. When the helium is gone, the core contracts, which triggers nuclear burning in shells above the core.

In stars at least eight times more massive than the Sun, carbon in the core begins to fuse when the helium is burnt out, with the reactions proceeding until the core is iron. Unprocessed carbon lies in a shell above the core. The end point of the star takes only a fraction of a second: the core undergoes catastrophic gravitational collapse. The seismic shockwave of the core collapse induces a cataclysmic chain of nuclear reactions that produce the elements heavier than iron. The star explodes as a supernova, leaving a remnant that is a neutron star or a black hole, with outer wreckage that slowly dissipates into interstellar space (Figure 2.6). Half of the carbon in Earth resulted from supernova explosions in our galaxy.

FIGURE 2.6 The Crab Nebula is the debris of a supernova explosion that was first observed on July 4, 1054. It is at a distance of 6500 light years.. *Source:* Hubble Space Telescope; NASA, ESA and Space Telescope Science Institute

For the source of the other half of Earth's carbon we need to look at the final stages of intermediate-mass stars, at about 1.6–6.0 solar masses. In the late phase of their evolution, such stars become red giants, in which carbon synthesis via the triple-alpha process takes place in a shell beneath the outer envelope of hydrogen. A spectacular transfer of energy by convection begins, dredging up carbon, nitrogen, oxygen and rock-forming elements from the deep. In the stellar envelope, the hydrodynamic flows produce intense stellar winds, in which the outer layers of the star are steadily driven into interstellar space as gas and dust grains.

Carbon-12 (^{12}C) is the more abundant of the two stable isotopes formed in stellar interiors, at 98.9 percent, all formed through the triple-alpha process. It is of particular importance in the laboratory because of its use as the standard by which the masses of all atomic nuclides are measured: its mass number is 12 by definition. The heavier isotope ^{13}C, with seven neutrons, accounts for only 1.1 percent of the natural abundance of the element. The triple-alpha process leads to ^{12}C building up in the stellar core, where capture of a proton yields ^{13}N, which loses a positron to form ^{13}C. The terrestrial ^{12}C/^{13}C ratio is 89, a little higher than the ratio of 85 for the Sun. For the inner solar system and meteorites, a ratio of 89–91 is the consensus value.[10] By contrast, in the nearby interstellar medium, the mean ratio value is 60 (with an uncertainty of ~15 percent).[11] What we learn from these ratios is that most of the local interstellar medium today has a greater ^{13}C abundance than the inner solar system, which condensed from the interstellar medium about 4.7 billion years ago.[12] Moderate chemical evolution of the local stellar neighborhood has taken place since the formation of the solar system. A central principle in recounting the history of carbon in the universe, the galaxy and on our planet is that chemical evolution takes place on every time and length scale in the universe.

This may be an appropriate point to anticipate some of the geochemistry that will feature later. We have just established that the two stable isotopes of carbon – ^{12}C and ^{13}C – are produced in two distinct stellar environments, and that the $^{12}C/^{13}C$ ratio can vary in both place (reservoir) and time. Isotope ratio analysis is a major tool of geochemistry for investigating the history of elements and their isotopes on and within Earth. We will encounter isotope ratios again when considering the cyclical processes involved in the distribution of carbon in Earth over deep time. But now we need to get back to cosmic carbon and the formation of the solar system.

⊙ GIANT MOLECULAR CLOUDS

By mass, the bulk chemical make-up of the interstellar medium is 70 percent hydrogen, 28 percent helium and 2 percent everything else. By number, only one atom in a thousand is heavier than helium. In the 13.7 billion years since the Big Bang, the stellar nucleosynthesis factories have added to the interstellar medium a 2 percent flavoring of 250 stable isotopes from lithium to lead. This is the gaseous medium into which the ashes of dead stars pour, where they face an onslaught of energetic processes that tears molecules apart as fast as they try to form. Hazards such as cosmic rays (protons and atomic nuclei moving at close to the speed of light), shock waves from exploding supernovas and ultraviolet radiation shatter the building of molecules. Turbulent motion and gravity's relentless attraction together produce structure on all scales in the interstellar medium. In regions of higher density in the midst of the galactic disk, molecular hydrogen and carbon dioxide are stable, allowing clouds to form. The largest of these – the giant molecular clouds – hold up to several million solar masses of gas and dust and are up to 150 light years in extent. Their temperatures are a cool 10–20 K (–263 to –253°C) and their dust blocks disruptive cosmic rays and ultraviolet radiation. Within this dark fastness, molecules form.

Where does cosmic carbon fit into this scenario? In the late 1960s, the new field of molecular radio astronomy commenced, using millimetre radio telescopes. Most molecules have the bulk of their rotational spectrum in millimetre wavelengths. Once radio astronomers had developed the technical capability to observe at those

FIGURE 2.7 Newborn stars are forming in the dusty pillars of the Eagle Nebula.
Source: Hubble Space Telescope; NASA, ESA and Space Telescope Science Institute

wavelengths, discoveries came thick and fast as telescopes scanned giant molecular clouds. In March 1969, radio astronomers used the 140-foot single-dish telescope at the National Radio Astronomy Observatory, Green Bank, West Virginia, to detect interstellar formaldehyde (H_2CO) in clouds at various locations in the galaxy. This pioneering discovery of "the first organic polyatomic molecule ever detected in the interstellar medium and its widespread distribution" suggested that the processes of chemical evolution in space were rather more complex than had been assumed.[13] Rival groups turned to molecular line astronomy in the search for new molecules. By the end of the century, the list interstellar molecules reached more than 100, including the organic molecules. When the discovery of the latter ethyl alcohol (ethanol) was announced in 1975, news media ran exciting accounts of how much liquor existed in space!

There is also a dust narrative for the cosmic carbon that arrived on Earth. In the regions of clouds where cryogenic temperatures pertain, CO, CO_2 and CH_4 may freeze onto dust particles. Mineral evolution in giant molecular clouds (Figure 2.7) probably begins when a giant star sheds its outer envelope through intense stellar winds or explodes as a supernova. Robert Hazen of the Carnegie Institution for Science, together with seven colleagues, has speculated that diamond was the first mineral to condense in carbon-rich zones at temperatures below 3700°C. Graphite came next, at 3200°C. The carbon-rich stellar envelopes would also have fostered the formation of carbides of iron and silicon. Let's now take a step back in time to discover how the mathematical models of the solar system came together from the eighteenth century and beyond.

⊙ FORMATION OF THE SOLAR SYSTEM

In the history of ideas, Pierre-Simon (marquis de) Laplace (1749–1827) was a brilliant savant who applied mathematical discipline rather than theological disputation to understand the origin of the solar system and our world. In time he became popularly known as "the French Newton." His excellence at mathematics shone out from the moment he commenced a degree of theology at the University of Caen, where he soon lost his faith, seeking in its place a new mission as a professional mathematician. In Paris, from 1771, he applied mathematics to statistics, physics and astronomy with spectacular success. Newton's theory of gravitation was the launch pad for Laplace's inquiries into the celestial mechanics of the solar system. In this endeavor he ditched Newton's attachment to divine intervention ("the finger of God") to account for the long-term stability of the solar system.

In 1795, Laplace won an excellent appointment as a professor of mathematics in Paris at the *École Normale*, an elite school newly established by the post-revolutionary French government. Despite his heavy teaching load, he found the time to give general lectures in which he used vivid language rather than dull mathematics. It was in one of these that he first outlined his "nebular hypothesis" for the origin of the solar system. Despite his mathematical prowess, Laplace presented his ideas in purely descriptive terms to make them accessible to a general audience. The lectures later formed the basis of a semi-popular book, *Exposition of the System of the World*, published in 1796. Laplace imagined that the planets began life within an extended cloud of gas slowly rotating about the Sun. Over time, the gas cloud cooled and contracted, rotating more rapidly as it did so and flattening into a disk. As the rotation rate increased, centrifugal force began to overpower the Sun's gravity, expelling rings of gas from the outer edge of the disk. Each of these rings later condensed into one of the planets we see today. Laplace insisted that any viable theory for the solar system must account for the features that we see today. In this respect, the nebular hypothesis appeared to be a success. It explained why the planets all orbit the Sun in the same direction and travel in nearly the same plane, and also why the planetary orbits are nearly circular.

When Laplace published his theory, he was apparently unaware that that Immanuel Kant (1724–1804) had proposed a nebular theory 41 years earlier. Why was that? In Kant's view, the solar system began as a cloud of material in space. Turbulent motions brought material together, allowing clumps to form, and these clumps attracted more material towards them. Denser clumps progressed more easily towards the Sun so that, when planets ultimately formed, the inner planets were denser than the outer ones. Like Laplace, Kant presented his ideas in words and did not attempt to support his case with mathematical reasoning. Sadly for Kant, his publisher went bankrupt almost immediately, and their stock, including his book, was seized. Kant's work vanished almost without trace. For many decades it was hardly known outside of Kant's home city of Königsberg in Prussia (now an enclave of Russia). It perhaps did not help matters that in his entire life Kant never ventured more than 15 kilometres (a three-hour walk) from his place of birth.

For many decades, the nebular hypothesis was the only scientific theory for the origin of the solar system, despite its lack of detail and absence of rigorous calculations. When new observations of certain nebulae by Herschel seemed to support his perspective, Laplace seized his opportunity. He cited them in the fourth and fifth editions (1813 and 1824) of his *Exposition*, which gave a longer version of the nebular hypothesis. Nevertheless, by the mid-nineteenth century, his hand-waving was flagged down by applied mathematicians, whose analyses showed that it simply would not work dynamically. Thereafter it languished for a century and a half, until new thinking and new observations overcame this neglect and slowly brought it back into prominence.

⊙ MAKING ROCKY PLANETS

For a while in the early twentieth century, theorists toyed with dynamic "catastrophe models" of planetary formation. They imagined circumstances such as the close approach to the Sun of some other star and how matter could be dragged out of the two through tidal forces. There were many variants on what catastrophic set-up could drag a filament of matter away from a star or the Sun and leading to the condensation of planets. By the mid-twentieth century, theorists had reluctantly concluded that stellar smash-ups or near misses, which had seemed to be on the right track for many decades, were in fact a road to nowhere.

So, they reimagined Laplace's hypothesis by envisaging that the primeval solar nebula condensed to form a disk, with the Sun at its center, *inside* an interstellar nebula, or cloud of dust and gas. Theorists now pondered how Earth and the planets might materialize in such a rotating disk. Towards the end of the 1960s, Viktor Safronov (1917–1999), a Russian physicist, produced a model according to which clumps of interstellar matter left stranded in the midplane of the solar disk gradually coalesced through a hierarchical process of accretion to form firstly grains, then pebbles, boulders next and finally planetesimals. These larger building blocks, measuring a few kilometres to several hundred kilometres across, then contracted under gravity's grasp to form protoplanets, which were the precursors of most of the current planets of the solar system.

Let's take a quick look at the evidence that planets self-assemble in circumstellar disks. In 1984, Beta Pictoris became the first star discovered to possess a bright disk of dust and debris. Since then, Hubble and ground-based telescopes have dissected the disk. It is easy to examine its structure (Figure 2.8) because it presents at an edge-on angle, and it is especially bright due to a very large amount of dust that scatters light from the central star. What's more, Beta

FIGURE 2.8 The circumstellar disk of gas and dust surrounding the star Beta Pictoris. This image was obtained in 2012 using the coronagraph of the Hubble Space Telescope to block light from the star itself..
Source: NASA, ESA and Space Telescope Science Institute

Pictoris is only 63 light years away, closer to Earth than most of the other known disk systems.

In 1995, photography by the Hubble Space Telescope yielded impressive evidence in support of Safronov's concept, when Jeff Hester and Paul Scowen of Arizona State University released a dramatic image of part of the Eagle Nebula (or M16, distance 7000 light years), exhibiting towering pillars of cold interstellar hydrogen gas and dust. New stars are forming here inside globules of gas and dust. In the mid-1990s, astronomer C. Robert Odell of Rice University, Houston, guided Hubble's camera to the bright Orion Nebula (M42, distance 1500 light years). In that prodigious star factory of luminous gases, he and colleagues spied the fuzzy blobs of thick disks of gas and dust surrounding new stars. These structures are infant solar systems, in which a central star is now pouring forth energy from the nuclear fusion of hydrogen in its core. This has huge consequences for the elements in the surrounding protoplanetary disk, including carbon and the other volatiles.

Close to the nascent Sun, temperatures were too high to allow the more abundant volatile substances in the nebula – those with comparatively low freezing temperatures, such as water, carbon diox ide and ammonia – to condense to their ices. The planetesimals that formed from the solid material present were therefore *deficient* in volatiles but *rich* in silicates and metals. Consolidation of rocky plan-etesimals built four small but dense inner planets – Mercury, Venus,

Earth and Mars. Further out beyond the interplanetary snowline, two gas giants – Jupiter and Saturn – and two ice giants – Uranus and Neptune – formed from the lighter elements and their molecular ices. Available evidence indicates that the asteroids, which orbit the Sun mainly in a belt between Mars and Jupiter, are remnants of rocky planetesimals that were prevented by Jupiter's gravity from consolidating into a planet at that location. In the violent end stage of the development of the embryonic Earth into a planet, huge collisions between the remaining planetesimals and the protoplanet swept Earth's orbit clear as the "feeding zone" spread ever wider. For 100,000 years or so, tumultuous impacts with mini-planets forged proto-Earth, a glowing blackened ball. Lava fountains from gigantic volcanoes and non-stop bombardment from planetesimal debris meteorites turned the surface into red-hot goo. Earth's earliest history of accretion from protoplanetary materials and the segregation of its core happened within 35 million years. Its internal structure had three nested domains: a very dense metallic core surrounded by a thick dense mantle and a thin crust at the surface.

The Earth's inner core consists primarily of nickel and iron. Heavy stuff sank to the bottom, thanks to gravitational attraction. A comparison with cosmic abundances indicates that the core is mainly iron with 5 percent nickel. The core is less dense than iron or iron–nickel alloys, which implies that around 10 percent is contributed by elements with a lower atomic number than iron (26) and with high cosmic abundance. In 1993, Bernard Wood concluded that sulfur and carbon were the most likely elements. Establishing the carbon content of the core was a major goal for the Deep Carbon Observatory.[14] Many siderophile (iron-loving) elements became sequestered in the core as solid solutions or in their molten state: gold, cobalt, iridium, osmium, silver and tungsten. There may also have been minerals such as iron carbide (Fe_7C_3) in the core by this stage. In the hot mantle, which is composed of silicates rich in iron and magnesium, the first crystals and minerals began to form at pressures hundreds of thousands of times greater than that of the atmosphere when the temperature had fallen to about 3000 K. The inner core most likely is a major reservoir for carbon in solution with iron.

⊙ THWACK! MOON MAKING

There was one final action shot in this cosmic drama about 100 million years after Earth's formation. There was a smash-up when a smallish planet drifted across Earth's orbit. Following the classical tradition for naming planetary bodies, this interloper was dubbed Theia, after the Greek goddess who gave birth to the Moon. Before the colossal sideswipe, Theia weighed in at maybe one-third of Earth's mass. It was in almost the same orbit as Earth, and so the two were destined to collide at some point. To demonstrate how the debris flung into space probably formed the Moon, theorists have run countless dramatic computer simulations of this "Big Thwack," but it continues as a work in progress.[15] The testimony of chemical isotopes in the 382 kg of lunar rocks brought back from six Apollo missions backs up the giant impact theory for the origin of the Earth–Moon system about 4.45–4.50 billion year ago.[16] The thermal energy released meant that both Earth and the Moon were covered in magma oceans that continued to take a thrashing from asteroids raining down. The unstoppable laws of thermodynamics meant Earth was inexorably cooling down, however. Solidification of the magma ocean probably took a million years or so. Earth now had its final form with core, mantle and crust.

REFERENCES

1. Oesch, P. A. *et al.* A remarkably luminous galaxy at z=11.1 measured with Hubble Space Telescope grism spectroscopy. *Astrophysical Journal* **819**, 129 (2016).
2. Prout, W. On the relation between the specific gravities of bodies in their gaseous state and the weights of their atoms (published anonymously). *Annals of Philosophy* **6**, 328 (1815).

3. Prout, W. Correction of a mistake in the essay on the relation between the specific gravities of bodies in their gaseous state and the weights of their atoms (published anonymously). *Annals of Philosophy* 7, 111–113 (1816).

4. Eddington, A. *Stars and Atoms* (Yale University Press, 1927).

5. Gamow, G. and Critchfield, C. L. *Theory of Atomic Nucleus and Nuclear Energy-Sources* (Clarendon Press, 1949).

6. Whaling, W. *Interview by Shelley Erwin* (American Institute of Physics, Oral History Archives, 1999).

7. Fowler, W. *Interview by Charles Weiner* (American Institute of Physics, Oral History Archives, 1973).

8. Burbidge, E. M., Burbidge, G. R., Fowler, W. A. and Hoyle, F. Synthesis of the elements in stars. *Reviews of Modern Physics* 29, 547–655 (1957).

9. Mitton, S. *Fred Hoyle: A Life in Science* (Aurum Press, 2005).

10. Woods, P. M. and Willacy, K. Carbon isotope fractionation in protoplanetary disks. *Astrophysical Journal* **693**, 1360 (2009).

11. Clayton, R. N. Oxygen isotopes in meteorites. In *Isotope Geochemistry: A Derivative of the Treatise on Geochemistry* (Academic Press, 2010), pp. 3–16.

12. Marty, B., Alexander, C. M. O. and Raymond, S. N. Primordial origins of Earth's carbon. *Reviews in Mineralogy and Geochemistry* 75, 149–181 (2013).

13. Snyder, L. E., Buhl, D., Zuckerman, B. and Palmer, P. Microwave detection of interstellar formaldehyde. *Physical Review Letters* 22, 679–681 (1969).

14. Wood, B. J., Li, J. and Shahar, A. Carbon in the core: its influence on properties of core and mantle. *Reviews in Mineralogy and Geochemistry* 75, 231–250 (2013).

15. Hazen, R. M. *The Story of Earth: The First 4.5 Billion Years, from Stardust to Living Planet* (Penguin Books, 2013).

16. Cameron, A. G. W. The origin of the Moon and the single impact hypothesis V. *Icarus* **126**, 126–137 (1997).

3   Deliveries of Cosmic Carbon Continue

⊙ ADDING THE VOLATILES, CARBON AND WATER

When the formation of the Moon and the phase of giant impacts had mostly run its course, a final round-up of the remaining planetesimals took place. Gravitational forces exerted by the eight planets scoured interplanetary space: some planetesimals were flipped into the Sun, others thumped into terrestrial or giant planets, and the remainder were expelled to interstellar space by gravitational slingshot effects. Unsurprisingly, gravity's purge of the solar system was not an entirely clean sweep. Comets, asteroids, meteors (meteorites once they make the journey through Earth's atmosphere) and mere specks of dust lingered. In our age, this debris has become an indispensable archival source of data for revealing how, over billions of years, the elements have been sieved and sorted for distribution throughout the solar system. Detailed investigations of the composition of asteroids and meteorites became a hot area of planetary science research at the beginning of the present century. In 2015, the NASA spacecraft *Dawn* began a three-year survey mission to Ceres, the largest asteroid.

Ceres had been discovered by accident in 1801, wandering in the enormous void of more than half a billion kilometres between the orbits of Mars and Jupiter. In 1772, the German astronomer Johann Bode (1747–1826) added a footnote to the second edition of an elementary text on astronomy in which he remarked on the almost regular geometrical progression in the sizes of the planetary orbits. Unlike others who were also aware of this pleasing pattern, in his footnote Bode wrote, "After Mars there follows a space ... in which no planet has yet been seen." The end of the eighteenth century saw astronomers perplexed by this yawning gap, located at one-third of the distance from Mars to Jupiter. On the first night of the nineteenth century, the puzzle was resolved by an unexpected discovery.

At Palermo Observatory, Sicily, Giuseppe Piazzi (1746–1826), the director, was at the telescope, updating a catalogue of star positions, when he noticed a "stellar object" that was not listed. On the following night he saw that it had moved relative to the background stars. The next day he penned a short letter to the *Journal de Paris* (the first French daily newspaper) announcing the discovery of a new comet. Although he initially thought it might be a comet, he changed his mind after a few months because he could not see a cometary tail.[1] Rather than finding a new comet, which was becoming a somewhat common achievement, he had discovered the missing planet between Mars and Jupiter. He gave it the name Ceres after the Roman goddess of agriculture. For the next half-century, Ceres was classified as a planet, but today it is officially a dwarf planet.[2]

Many similar but smaller objects were soon discovered lurking in the zone between Mars and Jupiter. Pallas was found in 1802, next was Juno in 1804, and three years later Vesta was found. As of 2018, there are more than three-quarters of a million minor planets (the preferred nomenclature). Automated searches of the asteroid belt will soon bring the census to a few million. Although the total amount of debris in the asteroid belt is only about 4 percent of the mass of the Moon, it is a fabulous resource of material from the earliest phase of the solar system. These minor planets are the closest to primordial samples of the solar nebula that we can get. For that reason, space agencies have launched reconnaissance missions to about a dozen asteroids and nine comets, aiming to discover what these objects can tell us about the final stages of the accretion of matter by Earth. NASA set a high priority on discovering the history of Earth's water and carbon. The fundamental research question to be settled is this: Did Earth's volatiles, including carbon and water, originate in the cold molecular cloud before the Sun's formation, or are they the products of a wholesale shuffling of elements and isotopes, dust and ices, comets and asteroids, which took place in the planet-forming disk?[3]

A recent example of such an asteroid project is NASA's *Dawn* mission, which launched in 2007 to study Ceres and Vesta, the two largest bodies in the asteroid belt. *Dawn* visited stony Vesta in 2011–2012 for a year-long orbital mission. Planetary scientists selected Vesta as a target because it is an asteroid like no other. There are meteorites in museum collections that were blasted off Vesta during collisions with neighboring asteroids early in the history of the solar system. They have a mineralogical make-up similar to Earth. When newly formed, Vesta soon had a rocky crust, a mantle and a core. The *Dawn* spacecraft confirmed the existence of its iron core, measuring about 220 kilometres across. From this we can deduce that Vesta completely melted in its early history, allowing its iron to sink to the core. *Dawn* moved on from Vesta to then spend three years investigating icy Ceres, where the mission highlight was the discovery of a global buried ocean of water ice (or briny mud) concealed below its dark and dusty surface. The discovery of copious supplies of subsurface water is highly significant because the origin of deep life in the solar system is contingent on liquid water being present in environments that are shielded from harsh cosmic radiation.

In 2005, the Japan Aerospace Exploration Agency's spacecraft *Hayabusa* managed to scrape dust from the surface of a near-Earth asteroid and return it to Earth. Researchers used electron microscopes and X-ray techniques to study the mineral chemistry of the dust particles, resulting in the confirmation of a long-standing suspicion that the most common meteorites found on Earth represent impact debris from stony asteroids.

Having scooped samples from the surface of an asteroid, the next target that motivated the space agencies was to land on a comet. Our oceans and the atmosphere exist because volatiles were still being added after Earth formed. One of the key goals of the European Space Agency (ESA) comet mission *Rosetta* was to chase, orbit around and land on the nucleus of a periodic comet. After a 10-year journey through the solar system, *Rosetta* arrived at comet

FIGURE 3.1 The double nucleus of Churyumov-Gerasimenko (also known as Comet 67P) was imaged in 2014 by the wide-angle camera on *Rosetta* from a distance of 15.5 kilometres. The larger lobe is about 4 kilometres in diameter.
Source: ESA/Rosetta/MPS for OSIRIS Team. Creative Commons

67P/Churyumov-Gerasimenko on August 6, 2014 (Figure 3.1). For two years it orbited the comet, gathering data on its environment and nucleus. Its lander *Philae* spent 64 hours sniffing the comet's gases and dust to determine their chemical compositions. The gases detected included water vapor and carbon monoxide and dioxide, along with traces of carbon-bearing organic molecules, such as formaldehyde. The dust grains had a suite of 16 carbon-rich and nitrogen-rich compounds. Some of the compounds detected by *Philae*'s instruments play a key role in the prebiotic synthesis of amino acids and sugars. These so-called ingredients of life form the backbone of the biomolecule RNA. It seems rather plausible then that comets, small asteroids and leftover planetesimals were responsible for the delivery of water and carbon (including organic matter) to Earth after the end of the giant impact phase.

⊙ CARBONACEOUS CHONDRITES CARRY CARBON

Meteorites conveniently deliver some of the most primitive solar system material available for study in our labs. They are our primary source of information about the early solar system because their chemical, isotopic and petrological features are a record of events that occurred in the first few millions of years of solar system history. Almost all meteorites are derived from collisions involving their parent bodies, which means they sample the highly diverse and numerous asteroids. Such samples are analyzed at a level of sophistication and detail that can only be achieved in a laboratory. This gives us access to priceless data on the earliest period of solar system formation, the mineralogy of the asteroid belt and the processes that have shaped asteroids over the life of the solar system. There are two distinct groups of meteorites with different histories: primitive meteorites (chondrites) and the differentiated meteorites (achondrites and stony-irons). We will focus on chondrites because they have a chemical and mineral composition that results from their formation processes in the solar nebula. Crucially, chondrites have not been modified by melting or differentiation of the parent body. By contrast, iron meteorites, consisting of an iron–nickel alloy, originate from the metal cores of planetesimals.

Chondrites are so named because they contain chondrules. These are small, round bodies that were once molten and are held in a matrix that is very fine grained and refractory. Ranging in size from a few micrometres to a centimetre, chondrules formed as molten drops before being accreted to their parent asteroids. These are the oldest solid materials in the solar system. Chondrites are *primitive* in the sense that the abundances of most of their chemical elements do not differ greatly from the solar system as a whole. About 5 percent of the meteorites that fall on Earth are classified as *carbonaceous* chondrites. As their designation suggests, their composition includes organic carbon compounds, such as amino acids, so they are of compelling interest in terms of the history of carbon on Earth and the origin of life.

In the early nineteenth century, the first scientific investigations of meteorites took place in France, where three remarkable meteorite falls transformed a field of skeptical inquiry into a new kind of science. We'll start in the lively market town of L'Aigle in rural Normandy, north-western France, where the population was almost 6000 according to the census of 1800. On March 26, 1803, a spectacular cascade of 3000 stones rained down from a clear blue sky. Many astonished citizens witnessed this fall. Two months later, the dramatic news finally reached Paris. Napoleon's minister for the interior, Jean-Antoine Chaptal (1756–1832), a brilliant multidisciplinary scientist, commissioned Jean-Baptiste Biot (1774–1862), a young professor of mathematical physics at the Collège de France, to make a full investigation. Accordingly, Biot undertook his inquiries through impressive work in the field. First, he checked out the local mineralogy and human artefacts. Then he wrote down witness statements from a diverse sample – travelers, coachmen, clergymen and citizen professionals – about the "rain of stones thrown up by the meteorite." Biot's report set out clearly the evidence that those fallen stones were of *extraterrestrial* origin, rather than caused by an atmospheric or volcanic phenomenon. He cited the physical evidence of the absence locally of any stone or artefact similar to the 37 kg of some 3000 fallen stones, as well as the testimonies of reliable eyewitnesses.[4]

The L'Aigle fall of an ordinary chondrite became a milestone in the development of our understanding of meteorites and their origins.[5] A turning point had actually been missed earlier, in 1794, when German physicist Ernst Chladni (1756–1827) published a 60-page pamphlet claiming that rocks had an extraterrestrial origin. He confined himself in the library at Göttingen, carrying out a literature search on reports of fireballs in ancient and modern sources. His largely deductive approach to these anomalous events led to his naive acceptance of any documented observation as pure fact. He concluded that meteors, fireballs and meteorites were different manifestations of the same phenomenon, namely the entry of a solid extraterrestrial object into the atmosphere. Although his hypothesis encountered much resistance, he kept on accumulating data from ancient sources until his death in 1827. By contrast, Biot's report enjoyed an enthusiastic reception because he had shown due diligence in writing down

statements from educated witnesses, thus avoiding the cliché of relying on the hearsay of *les paysans*. The L'Aigle meteorite is an ordinary
chondrite, the commonest class of meteorite.

The next big story in meteoritic science and deep carbon broke
on the March 15, 1806. Towards dusk, a few peasants toiling in fields
near to Alais in southern France (now Alès, Gard, in Occitanie) were
thunderstruck by a "great vividness seeming to be made by a cannon
shot which was preceded by a terrible rumble of thunder." Two
stones, one of 4 kg and the other of 2 kg, crashed from the sky. They
were so friable and black in color (Figure 3.2) that several of the locals
wondered if they were carbonized peat. Louis Jacques Thénard
(1777–1857), the professor of chemistry at the Collège de France,
published a chemical analysis of Alais (the official name of the
meteorite) later that year.[6] He found carbon at 2.5 percent by weight,
along with magnesium, iron oxides and nickel. This high carbon
content marked it out as an entirely new kind of space rock. Just
how different it was became clear in 1834. By chemical analysis, the
distinguished Swedish chemist Jons Jacob Berzelius (1779–1848) identified water in a meteorite for the first time. Probing further, he
isolated a blackish mess that was 12 percent elemental carbon. That

FIGURE 3.2 This fragment of the fragile Alais meteorite, about 20 mm in
diameter, is displayed in the Muséum National d'Histoire
Naturelle, Paris.
Source: Simon Mitton

was completely off the scale compared to the chemical make-up of most meteorites. This was a great moment for the history of deep carbon science: the recognition of carbon compounds in a unique meteorite. To his enduring credit as a careful scientist, Berzelius concluded that the presence of organics in the meteorite could not justify the conclusion that the organics must have been present at the site of its formation.[7] In other words, they were not a signature of life beyond Earth. He continued to speculate that meteorites were ejecta from volcanoes on the Moon, another of the popular hypotheses at the time.

Alais was the first meteorite to be recognized as a carbonaceous chondrite, a distinctive class: relative to ordinary chondrites, just 3.8 percent are classed as carbonaceous. Superficially, they are deceptively similar to charcoal briquettes. They are frangible, highly porous and brimming with soluble minerals: six months in the open in a wet climate leads to their disintegration. In terms of scientific discovery, that's why the L'Aigle meteorite was so important: it's how petrologists and chemists learned that meteorites come from outer space rather than thin air. Speaking of which, at that time, the noun *aerolite* – meaning air-stone – was still in common use. Ordinary chondrites are deficient in volatiles because they were formed in a high-temperature environment. Carbonaceous chondrites contain water up to 20 percent by mass as mineral hydrates and are generally richer in volatiles overall.

France turned up trumps for the third time on May 14, 1864, when a spectacular fireball as bright as the full Moon was observed all over the western part of the country. Twenty fragments from this exploding bolide crashed into the hamlet of Orgueil, 54 kilometres north of Toulouse (Figure 3.3). On this occasion, the news spread rapidly by mail. The French Academy of Sciences acted with alacrity: two letters reporting the observations were read at the Academy just a couple of days later. The first newspaper coverage appeared on May 17 in Montauban, the city nearest to Orgueil. It was a local account given by one Peyridieu, a retired professor of natural philosophy at

Fig. 3. Chute du bolide du 14 mai 1864.

FIGURE 3.3 This vivid record of the fall of the Orgueil meteorite was published in 1865 in an annual summary of scientific news, *Le triple almanach de Mathieu de la Drôme.*

Toulouse. He described a stone he had picked up that was about the size of an orange. This handy specimen was fragile, with a strong varnish. When he plunged it into a bucket of water, perhaps wanting to measure its density, it dissolved into mud as black as shoeshine![8]

Chemists and mineralogists in Paris and Toulouse quickly implemented a network approach to investigate this fall. The many witness reports generated intense activity, and the professionals were spurred on to act decisively by reports of the fragility of the samples. At the time, Gabriel Auguste Daubrée (1814–1896) was France's leading expert on meteorites. He was appointed professor of geology at the National Museum of Natural History in 1861. As curator of meteorite collections, he hit on the bright idea of augmenting the collection by requesting specimens (Figure 3.4) from a large number of correspondents. The resulting gifts enabled him to publish a catalogue of 86 distinct falls. He received numerous witness reports about the Orgueil fall from all over western France. The same was true for the famous head of the Paris Observatory and discoverer of Neptune,

FIGURE 3.4 Large fragment of
the Orgueil meteorite.
Muséum National d'Histoire
Naturelle, Paris.
Source: Simon Mitton

Urbain Le Verrier (1811–1877). Both Daubrée and Le Verrier received
many letters from reliable witnesses: a professor at Bordeaux, the
chief mining engineer at Bordeaux and the bishop and a mathematics
teacher at Montauban. Daubrée read these before the Academy on
May 23 and June 13, and they were published in *Comptes-Rendu*.

Bolstered by the sheer weight of evidence concerning the fresh-
ness of the fall, Daubrée and his associates then formed a consortium
to study the properties of the new meteorite. Specimens were distributed and tasks allocated. Drawing on his recent cataloguing experience, Daubrée classified Orgueil with Alais and two other
carbonaceous chondrites. He informed the Academy on May 30 that
it is "not only tender and friable, but that it disintegrates into an
impalpable dust as soon as it is put in contact with water." At the
same meeting, the chemist Stanislas Cloëz (1817–1883) announced
the first chemical analysis, reporting the presence of salts at 5.30
percent and a carbon content of 5.92 percent.[9] Five weeks later, he
gave a fuller account of the carbonaceous substances that he had
extracted with boiling hydrochloric acid. He noted that the insoluble
organic matter "looks very much like humus of some earthy
carburant ... analogous to peat or lignite."[10]

While the academicians busied themselves in Paris, the
Toulouse branch of this meteoritic carbon science network likewise

engaged in research. Pleasingly, their results complemented those of their Parisian colleagues. Two of the Toulouse scientists added that "had they put together all the stones they were presented they would have made the load of a donkey," which is surely a curious unit of measurement to be using 70 years after the adoption of the metric system! The "donkey work" would have amounted to 50 kg, according to a recent historical review of the Orgueil meteorite.[8]

In 1867, Daubrée published the definitive review of the scientific investigation of the meteorite.[11] He said that the richness of organic matter was a prominent property of the Orgueil stones, and cautiously remarked that they offer "a carbonaceous combination in a planetary body, where nothing proves so far the existence of organized beings, animal or vegetal." He eliminated the old hypothesis of a lunar volcanic origin for the meteorites. Daubrée estimated the mass of stones recovered for scientific analysis as 15 kg. Today, the surviving mass of stones from the fall is 14 kg, located in different museums, the main ones being Paris, Montauban, Toulouse, Prague and New York. For almost a century, what remained of the fragile meteorite lay in museum drawers, until March 16, 1961, when three chemists sounded a great alert to the chemistry and geology sections of the New York Academy of Sciences.

Bartholomew Nagy and Douglass Hennesey from Fordham University, together with Warren Meinschein of Esso Research, announced the discovery of biogenic hydrocarbons in a sample of the Orgueil meteorite from the American Museum of Natural History, New York. They stated that the hydrocarbons in the meteorite resemble in many aspects "the hydrocarbons in the products of living things and sediments on Earth ... the composition of the hydrocarbons in the Orgueil meteorite provide[s] evidence for biogenic activity." A staggering claim, indeed, particularly as all three were new to meteorite research.[12] Eight months later, Nagy was in the news again, together with George Claus, a microbiologist at New York University, as he had found "microscopic-sized particles, resembling fossil algae" in two carbonaceous chondrites, Orgueil and Ivuna (which had fallen in Tanzania in 1938). Orgueil was said to contain a few million "organized elements" per cubic centimetre, some of which had been observed "to undergo cell division."[13] Such claims about the carbon in a meteorite attracted absolutely sensational worldwide coverage. In France, *Planète*, a tremendously successful consumer magazine that served an exotic cocktail of news stories on the paranormal, ancient civilizations and pseudoscience, presented a sketch of one of the "organized elements," captioned: "FROM ELSEWHERE. ... This is the first image of an extraterrestrial being."

For a while, the many skeptical scientists suspended judgement, given the magnitude of the claim. Most notably, Nobel Laureate Harold Urey (1893–1981), who had discovered deuterium in 1932 and was an early pioneer of cosmochemistry, was sympathetic to the concept of extraterrestrial life, although he was not fully convinced that Nagy had found it. Urey's hypothesis was that the meteorites were of lunar origin. He speculated that early in its history (soon after the Big Thwack) the Moon had been contaminated with water and life-forms from Earth, which had then been returned courtesy of the meteorites. This incorrect idea later led to NASA managing risk by imposing quarantine on the Apollo samples, the eventual study of which falsified the hypothesis that carbonaceous chondrites came from the Moon.

The Orgueil meteorite came under intense scrutiny like no other meteorite before. The most critical researchers were Ed Anders and his team at the University of Chicago. They confirmed that the presence of "organized elements" that were nothing more magnetite grains (Fe_3O_4) plus pollen grains and various terrestrial contaminants.[14] Nagy did not accept this interpretation, so Anders and his team doubled down with even more intensive research. In 1964, they uncovered a hoax, probably perpetrated in the nineteenth century: someone had tampered with a sample distributed by the Musée d'Histoire Naturelle in Montauban, which had seed particles and pollen grains glued onto the stone![15] The discovery of this deliberate contamination cast doubt on the reality of extraterrestrial life.

Furthermore, it emphasized the importance of the highest levels of security in the supply chain from field to laboratory, which was a good outcome. The greatest care was thus taken when the next carbonaceous chondrite similar to Orgueil blazed to a fiery end across Australia.

⊙ ORGANIC RICHES IN THE MURCHISON METEORITE

Two months after the *Apollo 11* Moon landing, on September 28, 1969, a bright orange fireball exploded near the small rural village of Murchison in Victoria, Australia. Some 100 kg of meteorite fragments were hurled over the surrounding countryside of orchards, vineyards and dairy farms. The local residents searched the fields and byways, gathering samples straight away, most of them in a relatively good state of preservation, which reduced the chance of contamination. Eyewitness reports commented on solvent smells from the stones, which led to the speculation that they might contain organics. The largest fragment tipped the scales at 2.5 kg. Two-thirds of the fall went to the USA, where planetary scientists had prepared for the return of samples from the *Apollo 11* mission. Their labs were up and running with the instrumentation and trained staff for checking the presence of organics in the lunar samples. The Field Museum of Natural History in Chicago purchased 51.6 kg from Australia, and the US National Museum of Natural History (Smithsonian) bought 200 stones (19.8 kg). Both institutions continue to be the major sources of specimens for research purposes.

At the NASA Ames Research Center, California, investigators of the Murchison meteorite discovered the first convincing evidence of amino acids of extraterrestrial origin.[16] Ever since, biologists and geochemists have studied amino acids in carbonaceous chondrites. The situation 50 years later is that more than 100 amino acids and bases have been identified in the Murchison meteorite. It is one of the most primitive meteorites that we have because its chemical composition has not been greatly altered by significant planetary processing on the asteroid.[17] Murchison is a cosmic time capsule that is so rich in organic materials – amounting to over 2 percent carbon by weight – that it has become a reference standard for extraterrestrial organic chemistry. There is no shortage of specimens from the meteorite available, so its organic structures and compounds have been extensively analyzed by very sensitive high-resolution instrumentation that can detect abundances of tens of parts per billion.

⊙ COSMIC CARBON CHEMISTRY

The rare carbonaceous meteorites contain a wide range of organics that are essential to life. Carbonaceous meteorites may have contributed the first prebiotic building blocks to our early planet. In 1992, Chris Chyba and Carl Sagan suggested in *Nature* that 3.8–4.5 billion years ago, during the period of heavy bombardment, extraterrestrial organic molecules arrived on Earth.[18]

Highly sensitive techniques reveal that these carbonaceous meteorites contain a diverse suite of organic compounds, including the subunits for constructing the macromolecules from which nucleic acids, proteins and cell membranes could be assembled. Meteorites may, therefore, have provided the molecular toolkit for the origin of life on Earth, and possibly elsewhere in the solar system. The prebiotic cargo carried by chondrites included the units for constructing the first self-replicating molecules.[19]

The rich diversity and complexity of organic matter found in meteorites has rapidly expanded our knowledge and understanding of the extreme environments from which the early solar system emerged and evolved. The extraordinary diversity of organics in the Murchison meteorite successively sampled and recorded the processes of carbon chemistry in three distinct environments: interstellar space and the epoch of the formation of dust and ice grains; the solar nebula and its protoplanetary disk; and the chemistry of the parent body (asteroid).

Here's how Mark A. Sephton concluded a review of 40 years of research on organics, published in 2002:

> All known life is based on organic compounds and water and both
> of these have been present in the carbonaceous chondrites. As a

result, these meteorites constitute a valuable "natural laboratory" of prebiotic chemical evolution. Analyses of the organic matter in meteorites can provide an inventory of the types of chemical reactions and organic products which could have been significant on the prebiotic Earth. This inventory substitutes our terrestrial record which has been obliterated by biological and geological activity.[20]

The delivery of organics and volatiles to the surface of Earth has continued for billions of years. The annual rate of continuing accretion of cosmic dust, debris from comets and micrometeorites has been estimated at 40,000 tonnes a year. The commonest component is dust from grinding in the asteroid belt.[21]

In the present century, the research interface between meteoritics and the origin of life in the solar system has become a thrilling area of astrobiology. There is no doubt that extraterrestrial bodies with abundant carbon have reached the Earth since its formation. These natural samples of abiotic organic chemistry offer pathways and boundary conditions for considering the environments that were favorable for biogenesis. Half a century ago, when we learned about extraterrestrial amino acids for the first time, the initial speculation – quite naturally – was that these units could, by processes then unknown, self-assemble into self-replicating macromolecules. A consensus soon emerged that life must have been an emergent process through which biogenic atoms and molecules built the complex associations we see in the simplest forms of life.[22] Laboratory chemists found that a large variety of extraterrestrial organic molecules with counterparts in the biosphere could be made abiotically in the presence of liquid water, a discovery of profound significance for the origin of extant life, and for exobiology. In due course, we will return to these issues in later chapters that chronicle the history of deep life below the surfaces of the continents and the floors of the ocean basins.

REFERENCES

1. Piazzi, G. *Letter to Friend Barnaba Oriani, Director of the Brera Observatory, Milan* (1801).
2. Cunningham, C. J., Marsden, B. G. and Orchiston, W. Giuseppe Piazzi: the controversial discovery and loss of Ceres in 1801. *Journal for the History of Astronomy* **42**, 283–306 (2011).
3. Cleeves, L. I. *et al.* The ancient heritage of water ice in the solar system. *Science* **345**, 1590–1593 (2014).
4. Biot, J. B. *Memoires de la classe des sciences mathématiques et physiques de l'Institut National de France.* **7** (1803).
5. Gounelle, M. The meteorite fall at L'Aigle and the Biot report: exploring the cradle of meteoritics. *Geological Society of London, Special Publications* **256**, 73–89 (2006).
6. Thénard, L. J. Analyse d'un aerolite tombé dans l'arrondisement d'Alais, le 15 mars 1803. *Annales de Chemie* **59**, 103–112 (1806).
7. Berzelius, J. J. Ueber Meteorsteine. *Annalen der Physik* **109**, 113–148 (1834).
8. Gounelle, M. and Zolensky, M. E. The Orgueil meteorite: 150 years of history. *Meteoritics and Planetary Science* **49**, 1769–1794 (2014).
9. Cloëz, S. Analyse chimique de la pierre météorique d'Orgueil. *Comptes Rendus de l'Academie des Sciences Paris* **59**, 37–40 (1864).
10. Cloëz, S. Note sur la composition chimique de la pierre météoritique d'Orgueil. *Comptes Rendus de l'Academie des Sciences Paris* **58**, 986–988 (1864).
11. Daubrée, G. A. Complément d'observations sur la chute de météorites qui a eu lieu le 14 mai 1864 aux environs d'Orgueil. (Tarn et Garonne). *Nouvelles Archives du Muséum d'Histoire Naturell* **3**, 1–19 (1867).
12. Nagy, B., Meinschein, W. G. and Hennessy, D. J. Mass spectroscopic analysis of the Orgueil meteorite: evidence for biogenic hydrocarbons. *Annals of the New York Academy of Sciences* **93**, 27–35 (1961).
13. Nagy, B., Claus, G. and Hennessy, D. J. Organic particles embedded in minerals in the Orgueil and Ivuna carbonaceous chondrites. *Nature* **193**, 1129 (1962).
14. Fitch, F., Schwarcz, H. P. and Anders, E. Organized elements in carbonaceous chondrites. *Nature* **193**, 1123 (1962).
15. Anders, E. *et al.* Contaminated meteorite. *Science* **146**, 1157–1161 (1964).
16. Lovering, J. F., Le Maitre, R. W. and Chappell, B. W. Murchison C2 carbonaceous chondrite and its inorganic composition. *Nature Physical Science* **230**, 18–20 (1971).

17. Rubin, A. E., Trigo-Rodríguez, J. M., Huber, H. and Wasson, J. Progressive aqueous alteration of CM carbonaceous chondrites. *Geochimica et Cosmochimica Acta* **71**, 2361–2382 (2007).
18. Chyba, C. and Sagan, C. Endogenous production, exogenous delivery and impact-shock synthesis of organic molecules: an inventory for the origins of life. *Nature* **355**, 125 (1992).
19. Callahan, M. P. *et al.* Carbonaceous meteorites contain a wide range of extra-terrestrial nucleobases. *Proceedings of the National Academy of Sciences* **108**, 13995–13998 (2011).
20. Sephton, M. A. Organic compounds in carbonaceous meteorite. *Natural Product Reports* **19**, 292–311 (2002).
21. Brownlee, D. E. The origin and properties of dust impacting the Earth. In *Accretion of Extraterrestrial Matter throughout Earth's History* (Springer, 2001), pp. 1–12.
22. Pizzarello, S. and Shock, E. The organic composition of carbonaceous meteorites: the evolutionary story ahead of biochemistry. *Cold Spring Harbor Perspectives in Biology* **2**, a002105 (2010).

4   On the Nature of Earth's Interior

Treating Earth as a system in which life plays an important role in controlling the environment is not new. In fact, speculation on how the physical and living elements of the Earth interact goes back to antiquity, and earlier, with deities capriciously intervening in the affairs of the human population. The notion of a single integrated system mediated by gods was the stuff of religion and legend. The Greek cosmologists dismissed arbitrary gods as controllers of the universe. Instead, they considered that the order of nature and the cosmos follows eternal cycles. Democritus (c. 460–370 BCE) proposed an atomist theory of matter that invoked an early form of the conservation of energy: atoms are in eternal motion. The Roman poet and philosopher Lucretius (96–55 BCE) drew on the atomic theory of Democritus to account for a variety of terrestrial phenomena, such as earthquakes and lightning, according to physical principles. Some Greek philosophers, including Thales (624–546 BCE), Anaximenes (585–528 BCE) and Heraclitus (535–475 BCE), taught a philosophical doctrine that there is a form of life in all matter. The rationalist philosophical styles of the ancient world were lost until the world of ideas was revived in Europe in the fifteenth century by the rise of humanism in Italy.

The first person to arrive at what we would call an evidence-based result in Earth science was the philosopher Eratosthenes of Cyrene (276–194 BCE), who came from an ancient Greek city on the north-west coast of present-day Libya. An all-round scholar, skilled in arithmetic, geometry and astronomy, in about 245 BCE he settled in Alexandria, on the invitation of Ptolemy III who appointed him as a

tutor. Around 240 BCE, Eratosthenes secured one of the most important academic posts in the ancient world when Ptolemy III appointed
him director of the Library of Alexandria. At about this time, he made
a surprisingly accurate calculation of Earth's circumference. Although
his own treatise about his measurement of Earth is lost, we have
sufficient detail from later writers to reconstruct his geometrical
method.

Eratosthenes was aware that on a midsummer's day the Sun is
directly overhead at Syene (Aswan) in Upper Egypt. In his home city of
Alexandria, a major port on the Mediterranean, which is more or less
due north of Syene, he measured the size of the small shadow cast at
noon by a vertical stick (or *gnomon*). The shadow's length told the
geometer the difference between the angular distance of the Sun at
Alexandria and Syene: it was one-fiftieth of a complete circle (7.2°).
For the length of the meridian arc between Alexandria and Syene he
adopted 5000 stadia (a suspiciously round number), which he already
knew from the camel caravans that plodded the major trade route
between the two cities. Multiplying by 50 gave the value of 250,000
stadia for Earth's circumference. Actually, he rounded that up to
252,000 stadia to get a value exactly divisible by 60: 700 stadia per
degree of latitude.

We don't know what length Eratosthenes used for the stadion.
How could we? Its value varied somewhat from state to state.
Scholars have variously concluded that he was almost spot on, to
having an error of about 15 percent. In assessing his achievement,
the accuracy is not that important, but rather his ingenuity and
boldness in deciding that a property of the Earth as a whole could be
found by using a method that combined geometry and measurement.
It was an amazing breakthrough, representing the first use of
geometry to estimate the size of the world. His successors who
applied geometry to measure the universe and its mechanisms were
no less than Claudius Ptolemy, Nicolaus Copernicus, Isaac Newton
and Albert Einstein. Eratosthenes is also famed as "the father of
geography." He commented on the ideas of the nature and origin of

Earth: he had thought of Earth as an immovable globe, while on its surface was a place that was changing. He hypothesized that at one time the Mediterranean had been a vast lake that covered the countries that surrounded it and had only become connected to the ocean to the west when a passage had opened up sometime in its history.

⊙ MINING DEEP EARTH BEGINS

Nearly two millennia elapsed before the next significant advance in Earth sciences. In continental Europe in the second half of the fifteenth century, a small number of scholars revolutionized the classical ways of understanding the world: the Renaissance polymath Leonardo da Vinci (1452–1519), the humanist scholar Erasmus of Rotterdam (1449–1536), the astronomer Nicolaus Copernicus (1473–1543), the theologian Martin Luther (1483–1546) and the physician Paracelsus (1493–1541). This handful of names is remembered today for the rejection of dogma and the promotion of imaginative approaches to thinking styles. In doing so, they laid the foundations for the seventeenth-century scientific revolution based on empiricism.

The year 1494 saw the birth of Georg Pawer (1494–1555; Georg Bauer in modern German) in Glauchau in Saxony.[1] As was the custom at the time, his teachers Latinized his birth name to Georgius Agricola (Figure 4.1). His birth was scarcely 40 years after the printer Johann Gutenberg (1400–1468) in Mainz first produced printed books. By the time Georgius was six years old, printers in dozens of cities had already run off 20 million books. At around the same time, Erasmus, who in the 1530s would become a strong supporter of Agricola, completed his higher education at the Collège de Montaigu of the University of Paris. Agricola, gifted with great intellect, adopted the new learning, becoming a humanist of exceptional ability with great enthusiasm.[2] He enrolled at the University of Leipzig (founded 1409) at the age of 20, gaining there his Bachelor of Arts degree. Then he taught Greek and Latin at a grammar school in Zwickau, where he became a rector (principal) in 1519 with special responsibility for

FIGURE 4.1 Bronze sculpture of Georgius Agricola by Rudolf Löhner in Glauchau, Saxony, his place of birth.
Source: Photographer Walter Möbius. © SLUB/Deutsche Fotothek/Walter Möbius. Creative Commons License

organizing a new school of Greek. After travelling and studying in Italy for about three years, he came to Saxony with a medical degree. In 1527, he became a physician in the booming mining town of Joachimsthal (now Jáchymov) in the midst of the most prolific metal-mining district of central Europe. Silver had been found in the area in 1470, and soon thousands of prospectors were allowed to settle freely. Every aspect of the ore-mining operations fascinated Agricola.

By his own admission, Agricola spent all of his time when not required for medical duties visiting mines and smelters, conversing with mining experts and studying classical texts on mining and minerals. His first book, published in 1530, is a minor treatise on the mining of ores, written as a dialogue between a miner/mineralogist and two learned persons, both of them scholars and physicians.[3] The dialogue's main theme is the correlation of minerals mentioned by the ancient authors with those found in the Saxon mines. The three conversationalists wander through the mines to debate on particular items that attract their curiosity. Although Agricola's first attempt at writing a text based on his own observations lacks systematic or logical arrangement, statements of historical or technical interest pop up here and there. For example, the book includes the first descriptions of several subsurface minerals, some of which had been known since ancient times. Dialogue was a significant ancient literary genre that the humanists of early modern Europe adopted. Through this narrative style, Agricola joined the good company of several influential writers of the sixteenth century. For modern readers, a famous example of the technique is found a century later in Galileo Galilei's *Dialogue on the Two World Systems* (1632).

Two of Agricola's truly groundbreaking works rolled off the printing press in 1546: *De ortu et causis subterraneorum* or "The order and causes of subterranean materials," which is the first textbook on physical geology, and *De natura fossilium* or "The nature of minerals," the first textbook of mineralogy. The Latin *fossilium* translates roughly as "anything dug up," by which Agricola meant earth, stones, marbles, metals and "congealed juices," among which he counted bitumen and other "natural carbons."[4] He organized minerals on the basis of their physical properties, such as color, density, odor, taste, shape and so on, rather than by name or any other system. He did so as a natural historian, informed by his own observations and interpretations of specimens he collected personally.

Although the bulk of Agricola's mineral collection was probably sourced from the rich silver mines of Saxony, he travelled widely to acquire specimens. In a customary dedication of the book, he informs his patron, the illustrious Maurice, Duke of Saxony (1521–1553), that:

> I have attempted to discuss those minerals not found in Germany but in other parts of Europe and certain parts of Asia and Africa. In the discussion of these minerals, learned men, traders and miners have been of great assistance to me.

His obsequious dedication concludes with this association:

> No one can examine this work carefully and not realize that my studies will become better known and of added value through the greatness of your fortune, honour and virtue.

Be Strong Oh Illustrious Prince
Chemnitz, February 15, 1546

This masterpiece alone would have secured his reputation today. However, he was already preparing an even greater work that would not be published in his lifetime: *De re metallica*, the famous classic on mining and metallurgy. His began planning this in the late 1520s, and although he had completed it by 1530, he delayed sending it to press until 1553. In Basle, the noted printer of academic books and friend of Erasmus, Hieronymus Froben (1501–1563), published it in 1556, just four months after Agricola died from a short fever. Agricola's approach combined a deep technical and financial knowledge of mining with an underlying interest in the health of mineworkers. The practical know-how included sections on prospecting, on mine construction and on operational matters. Two hundred and seventy large woodcut illustrations enhanced the volume, but cutting them was so time-consuming that publication was delayed for years. Agricola assumed direct responsibility for the quality of the illustrations: he sought out professional illustrators to make the sketches and stood over the craftsmen block-cutters. The resulting beautiful woodcuts (Figure 4.2) depict a variety of mining and refinery processes, including haulage of ores, mine ventilation and drainage and the stoking of furnaces.

In the history of printed books, *De re metallica* counts as the first highly illustrated textbook of technology. Within it, Agricola reviews the knowledge that generations of miners had accumulated, thereby promoting their remarkable achievement through its detailed and intelligent exposure. He even commented on the waste of resources in many of the mines around Freiberg and the Harz Mountains, suggesting that they would cause problems for future generations.[5] *De re metallica* became the chief handbook on mining and metallurgy for several succeeding generations, circulating widely in Europe throughout the sixteenth and seventeenth centuries. Going through at least 10 editions in 3 languages, it advanced the sciences of

FIGURE 4.2 Woodcut of silver mining in Saxony. Agricola, *De re metallica*, 1556.

geology, mineralogy and mining engineering for 180 years until it was superseded in 1738 by the appearance of Christoph Andreas Schlüter's (1668–1743) metallurgical handbook. The classical Latin of Agricola's text ensured its wide circulation among educated Europeans, although it would set a major challenge for future translators because he had to invent the Latin versions for several hundred expressions in Mediaeval German that dealt with mining and milling terminology (Figure 4.3).

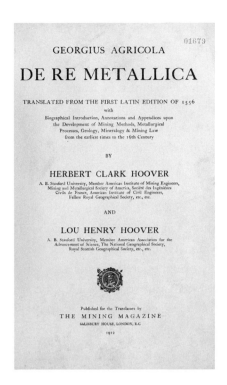

GEORGIUS AGRICOLA

01679

DE RE METALLICA

TRANSLATED FROM THE FIRST LATIN EDITION OF 1556
with
Biographical Introduction, Annotations and Appendices upon
the Development of Mining Methods, Metallurgical
Processes, Geology, Mineralogy & Mining Law
from the earliest times to the 16th Century

BY

HERBERT CLARK HOOVER

A. B. Stanford University, Member American Institute of Mining Engineers,
Mining and Metallurgical Society of America, Société des Ingénieurs
Civils de France, American Institute of Civil Engineers,
Fellow Royal Geographical Society, etc., etc.

AND

LOU HENRY HOOVER

A. B. Stanford University, Member American Association for the
Advancement of Science, The National Geographical Society,
Royal Scottish Geographical Society, etc., etc.

Published for the Translators by
THE MINING MAGAZINE
SALISBURY HOUSE, LONDON, E.C.
1912

FIGURE 4.3 Title page of the 1912 London edition of Georgius Agricola's *De re metallica*. Translated from the first Latin edition of 1556 by Herbert Clark Hoover and Lou Henry Hoover. Herbert Hoover amassed a fortune in the gold fields of Australia and China, working as a highly successful mining engineer. He served as the 31st President of the USA.

Agricola's lasting contribution to the history of deep Earth science arose from his decisive breakaway from the philosophical style of the ancient world. The learned world into which he was born had continued to turn to the Greeks and the mediaeval alchemists for fundamental explanations of natural phenomena. He had benefitted from the most thorough education of his time, gaining an exhaustive knowledge of classical literature as well as of medicine and the sciences. He corresponded with the leading scholars of Europe. In refuting the ancient teachings of the Peripatetics, he mitigated the risk of ridicule by asserting that the results of careful observation trumped the inductive speculation of the past. Here's how he justified his methods:

> Those things which we see with our eyes and understand by means of our senses are more clearly to be demonstrated than if learned by means of reasoning.[6]

He was a true pioneer in building the foundations of the scientific method on deduction from observed phenomena. I would place Agricola alongside Copernicus in terms of impact on science. Both occupy formidable places in the awakening of learning in the sixteenth century. Both replaced baseless speculation with applied research and observation. In terms of the acceptability of their ideas, Agricola had the advantage that observation is easier in the Earth sciences – besides which, he did not produce hypotheses likely to offend the Church. Even after 1517, the year when Luther had sparked the Protestant Reformation in Saxony, theologians in Rome had no interest in Earth's interior: they had no intention of going to that hellish place. We can surely count Agricola as a pioneer in the science of deep Earth. His views on the origin of ore deposits are particularly striking because they resonate with modern views. He proposed that ore bodies are deposits from the circulation of ground waters: in our terminology, this is the process of ore genesis by the movement of hydrothermal water within the crust.

When we confront the richness of today's deep geosciences, we could flippantly regard Agricola (the farmer!) as a lightweight for having merely scratched the surface while fossicking for precious metals, but that would be a grave error, similar to dismissing the work of Copernicus because he considered only the solar system while neglecting the vast universe beyond. For Agricola, the extraction of minerals (chiefly metals) did not differ from agriculture. Metals had been created to support human life. Agriculture, a surface activity, was dependent on the extraction of metals from the deep for its tools. Accordingly, in *De re metallica*, he reflects that:

> If there were no metals men would pass a horrible and wretched existence in the midst of wild beasts; they would return to the acorns, and fruits, and berries of the forest. They would feed upon the herbs and roots they plucked up with their nails. They would dig out caves in which to lie down at night, and by day they would rove in the woods and plains at random like beasts.

☉ EARTH'S MAGNETIC ATTRACTION

In the early modern period (c. 1500–1800), the rational consideration of Earth's interior began with William Gilbert (1544–1603), an experimentalist who made the first advances in understanding the properties of magnets and who speculated on the origin of Earth's magnetic field. Gilbert's primary scientific treatise, *De re magnete*, published in Latin in 1600, is a milestone in the history of science. Gilbert used spherical loadstones and explored how tiny magnetic needles, free to turn, set themselves at different points on such a stone. Gilbert concluded that the solid Earth behaved as a giant loadstone, and that the source of its magnetic force lay within Earth (Figure 4.4). Through his treatise, Gilbert became the founder of the new field of terrestrial magnetism. In the book, he advised that natural philosophers should "look for knowledge not in books but in things themselves."

In 1600, the Royal College of Physicians appointed Gilbert to their Presidency, and the following year he became court physician to Queen Elizabeth. During her long reign, Britain became a major seafaring nation, trading internationally from ports in London and the Thames Estuary. Queen Elizabeth had chartered a powerful group of London merchants to form the East India Company for promoting England's trading links with East Asia. To assist with long-haul

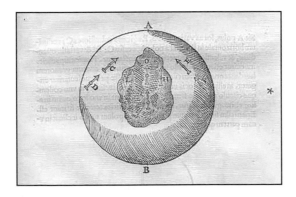

FIGURE 4.4 Gilbert's simple loadstone model for the source of the geomagnetic field.

navigation, sailors relied on the mariner's compass, in which a horizontal magnetic needle swings on a vertical pivot. This handy device, already perfected in Italy by the thirteenth century, enabled mariners conducting international trade to trace their position relative to Earth's magnetic field. Thanks to this, trade expanded, which in turn aided the rise of the Italian city states. At the turn of the thirteenth century, intrepid Genoese explorers ventured into the Atlantic, following the coast of Africa. The biggest challenge mariners faced when using the compass for long-haul trips was what we now call magnetic variation: the angular difference in the direction the compass needle points to compared to true north. On the Mediterranean Sea the difference is small, but its variation across the vast Atlantic Ocean is considerable. This was first noted in 1492 by Christopher Columbus, who kept the discovery secret for fear of panicking his crew.

⊙ EDMOND HALLEY INVESTIGATES GEOMAGNETISM

Edmond Halley (1656–1741) was the natural philosopher who investigated the cause of magnetic variation. He became curious about the magnetic compass while at St. Paul's School in the City of London. Excelling in classics and mathematics, Halley's curriculum also included astronomy and navigation. It was here that Halley made his first measurement of magnetic variation. Then, at the age of 20, he suddenly dropped his studies at Oxford on being granted free passage by the East India Company to the volcanic island of St. Helena, which had become an important replenishment stop for the spice trade ships of the East India Company that governed the territory. The Atlantic voyage offered Halley plenty of opportunities to measure magnetic variation. At the southernmost outpost of the English realm, Halley set up an astronomical observatory, where he made the first survey and catalogue of the stars in the southern hemisphere. When he returned to The Queen's College, Oxford, he immediately set to work on his star catalogue, which the Royal Society published in November 1678, the same year that he was elected to the Fellowship of the Society. There was some doubt about his academic standing, however, because he had not satisfied the rules on residency for a Master of Arts degree. King Charles II duly obliged by signing a letter authorizing the award of such a degree on the strength of Halley's astronomical observations at St. Helena. According to his biographers John and Mary Gribbin, "Halley received the first degree ever awarded specifically for research."[7]

Halley dipped in and out of studies on magnetism and compass variation, producing in 1683 the first of several papers on geomagnetic phenomena.[8] First, he compiled a table of the variations from all parts of the globe, listing "mostly the observations of persons of good skill and integrity" made in the seventeenth century. By trial and error, he worked out a configuration of internal magnets that pretty much agreed with the data.

> From Study in this difficult subject, and believing that I have put it past doubt, that there are in the Earth Four such Magnetical Points or Poles which occasion the great variety and seeming irregularity which is observed in the variations of the Compass.

During the middle of his career, he continued to ponder this intellectual puzzle. In 1692, Halley set out his ideas on Earth's internal structure.[9] To explain the magnetic variations, he proposed that Earth has an outer shell 800 kilometres thick enclosing an inner core, separated by an air gap. In this way, the outer shell and inner core could each have their own magnetic field, with two poles. The magnetic variations could be the result of different rotation rates of the shell and inner core. He "adventured" further in developing the scheme by having subdivisions of the interior, positioning two inner shells concentrically about a solid core. Anticipating a hostile reception to this unnatural scheme, he delicately admitted:

> We have adventured to make the Earth hollow and place another Globe within it: and I doubt not that this will find Opposers enough. I know 'twill be objected, that there is no Instance in Nature of such a thing.

Like Gilbert but more so, Halley made an immense contribution to the history of ideas about the nature of Earth's interior. In his surveys of southern stars and his maps of magnetic variation he stressed the importance of reliable data as the right way to test hypotheses. Notably, his approach to natural philosophy involved him getting his own measurements, then reducing those measurements systematically and presenting the results in the *Philosophical Transactions*, thereby making them freely available to all scholars. Commenting that more observations would be needed in the future for a "nice determination ... of several other particulars in the Magnetick Systems," he hit on a brilliant idea: crowdsourcing!

> All that we can hope to do is to leave behind us Observations that may be confided in, and to propose Hypotheses which after Ages may examine, amend or refute. Only here I must take leave to recommend to all Masters of Ships and others, Lovers of natural Truths, that they use their utmost diligence to make, or procure to be made, Observations of these variations in all parts of the World, as well in the North as well as the South Latitude (after the laudable custom of our East-India Commanders), and that they please communicate them to the Royal Society.

Halley was the first to curate good-quality geo-*physical* data on the geo-*magnetic* field and the first to propose a clockwork model of how Earth's surface magnetism might arise naturally in the deep interior. His global call to Masters of Ships is the first example of a natural philosopher attempting to crowdsource geophysical data. His appeal even included precise instructions on the need to avoid errors caused by refraction when measuring the Sun's amplitude (the arc along the horizon of the rising point and due east), an essential step in finding the angle of magnetic variation.

Halley's paper of 1692 marks an important development of the scientific method. From his long sea voyages, he had understood the problems encountered when conducting observations on a global scale in a consistent and coordinated manner. His solution – to arrange for many reliable observers to send results to a hub (a data center, as we would say), where they would be analyzed by experts – began to emerge in the field of seismology two centuries later. The International Polar Years (1882–1883 and 1932–1933) used coordinated scientific observation of the polar regions to advance several fields, including geophysics. And then the International Geophysical Year (1957–1958) took the networking of data acquisition and sharing of resources to an entirely new level. The Deep Carbon Observatory's mission, too, is a global research activity in the geosciences being carried out by actions that break down the traditional boundaries of disciplines to forge a new interdisciplinary field of carbon science.

⊙ HENRY CAVENDISH WEIGHS EARTH

In the two centuries after Halley's 1692 paper, scientific progress on understanding the nature of Earth's interior was slight. The natural philosopher Henry Cavendish (1731–1810), well known for the precision of his experiments on the chemical composition of air, conducted a famous experiment to weigh Earth. Henry's family traced its lineage back eight centuries to Norman times, and they were connected to the highest levels of nobility in England: two of his grandparents were dukes. From the late 1780s, in addition to his townhouse with his library, he had a fine mansion south of London in Clapham – then a small village – for his instruments and experiments. In a hut on the Common, he set up a torsion balance with which he measured the tiny deflection due to the gravitational attraction that two massive lead spheres (158 kg) exerted on two small lead balls (0.73 kg). Since the small spheres were attracted vertically by the mass of Earth and horizontally by the masses of the large spheres, he could derive the mass of Earth from the angle of deflection. His result, published in 1798, gave a mean density 5.448 ± 0.033 times greater than that of water, within 1.2 percent of the modern value of 5.514. The Cavendish experiment confirmed Newton's hunch that stuff heavier than the surface rocks had sunk and "made for the center." We shall return to the topic of the composition and fractionation of Earth's core in later chapters. The Cavendish experiment brings us to the end of the eighteenth century, as well as the end of the early modern period. The timeline of our journey of discovering the science of Earth's interior now moves to the nineteenth century, a period when geology matured as a physical sciences discipline and as a profession.

REFERENCES

1. Hoover, H. and Hoover, L. Introduction. In English Translation *De re metallica* (The Mining Magazine, 1912).
2. Hannaway, O. Georgius Agricola as humanist. *Journal of the History of Ideas* **53**, 553–560 (1992).
3. Agricola, G. *Bermannus, sive de re metallica* (Frobenianis, 1530).
4. White, G. W. *De natura fossilium* (*Textbook of Mineralogy*). Georgius Agricola. *Journal of Geology* **65**, 113–114 (1957).
5. Bowell, R. J. What is sustainability in the context of mineral deposits? *Elements* **13**, 297–298 (2017).
6. Agricola, G. *De ortu et causis subterraneorum Book III* (Frobenianis, 1546).
7. Gribbin, J. and Gribbin, M. *Out of the Shadow of a Giant: How Newton Stood on the Shoulders of Hooke and Halley* (HarperCollins, 2017).
8. Halley, E. A theory of the variation of the magnetical compass. *Philosophical Transactions of the Royal Society, London* 13, 308–312 (1683).
9. Halley, E. An account of the cause of the change of the variation of the magnetick needle, with an hypothesis of the structure of the internal parts of the Earth. *Philosophical Transactions of the Royal Society, London* 17, 563–578 (1692).

5   Earth's Physical Interior Revealed

◉ "THE MOST INGENIOUS BOOK I EVER READ"

For many centuries, the search for concordances with ancient texts dominated the timeline of natural history, and there was no connection with geology. This scenario did not appeal to Robert Hooke (1635–1703) who, in 1662 at the age of 27, was appointed the curator of experiments at the Royal Society. He became a great polymath, Renaissance man and indeed a jack of all trades: architecture, astronomy, biology, mechanics, optics and geology came within his remit. His numerous inventions included the iris diaphragm, the universal joint and the compound microscope. He accumulated a huge personal fortune (£8000 in gold and cash, equivalent to a few million pounds today) when he and his friend Christopher Wren worked intensively in 1666–1680 as the principal architects rebuilding London after the Great Fire. Hooke's career as curator at the Royal Society established his reputation as the first natural philosopher to focus on experiments made with precision instruments. For him, Nature was a great machine or engine in motion, the secrets of which could be uncovered with ingenious devices. In 1665, Hooke completed *Micrographia*, describing his adventures of discovery in the inner world revealed by his microscope. The contents of this book, one of the most significant scientific textbooks ever written, were sensational. Published by the Royal Society, it was the first illustrated book on microscopy, and it became a bestseller, inspiring wide public interest in the new science. The diarist Samuel Pepys read the book voraciously and recorded:

> Before I went to bed I sat up till two o'clock in my chamber reading Mr Hooke's Microscopical Observations, the most ingenious book I ever read in my life.

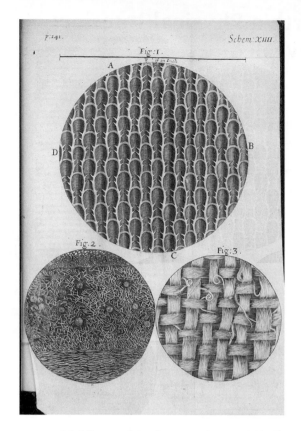

FIGURE 5.1 *Micrographia*, plate opposite page 141. The scale bar at the top indicates 1 inch (25 mm). 1: Surface of seaweed; 2: rosemary leaf; 3: fine linen. Hooke noted that he "could exceedingly plainly perceive it all perforated and porous, much like Honey-comb, but that the pores of it were not regular ... these pores, or cells ... were indeed the first microscopical pores I ever saw."

Micrographia has a remarkably detailed description of plant cells, the term "cells" being his coinage (Figure 5.1).

Hooke's fascination with fossils displayed his talents to maximum advantage over a period of 30 years. During his childhood, he had developed this interest through his finds from the limestone coast of the Isle of Wight in southern England, where he was born. He declined to consider fossils as anything other than the petrified remains of organic creatures. He argued:

> Nature does nothing in vain; it seems, I say, contrary to that great Wisdom of Nature, that these prettily shaped bodies should have all those curious Figures and contrivances ... generated or wrought by a Plastick virtue, for no higher end then only to exhibit such a form.

In other words, unlike his colleagues, he steadfastly refused to invoke the Deluge and its drowned world to account for marine fossils that lay in rocks far from the sea. Instead, he concluded that Earth's surface is a wreckage of mighty upheavals caused by cataclysmic earthquakes driven by inner heat.

Almost two centuries before Charles Darwin (1809–1882) would do likewise, Hooke realized that the fossil record has captured changes in the organisms living on our planet. He suggested that many fossils represent "Species of Creatures in former Ages, of which we can find none at present." Through his observations of fossils, Hooke became confident that Earth was much older than the 6000 years that Archbishop James Ussher (1581–1656) had announced in 1650. Hooke's philosophical approach to the immensity of geological time allowed free range to his genius. Thanks to an indefinite timescale, he perceived the mechanisms of change (the weathering of rocks) as operating cyclically on Earth. No vestiges of the original rocks could have survived "the vicissitudes of changes."

⊙ STRATIGRAPHY: A LAYERED APPROACH TO EARTH HISTORY

The field of stratigraphy began in a work of 1669 by Nicholas Steno (1638–1686), a Danish pioneer of anatomy and geology. Steno outlined four basic principles to interpreting layers of sedimentary rocks. These principles are: *superposition*, which states that younger rocks generally overlie older rocks, even if geological upheavals tilted them afterwards; *lateral horizontality*, which states that the sedimentary layers at the time of their formation lay parallel to the horizon; the *cross-cutting relationship*, which states that rock layers must be older than any intrusion; and *lateral continuity*, which states that material

FIGURE 5.2 James Hutton. From a portrait by Sir Henry Raeburn. National Galleries, Scotland.

forming a stratum spreads continuously over the surface of Earth unless blocked by other solid bodies.

Steno's principles and the gradual development of geological fieldwork inspired the Scottish geologist James Hutton (1726–1797) (Figure 5.2). He laid out his thesis that the present land and strata are not original, but rather were "formed by the operation of secondary causes." Hutton's innovative 1785 theory visualized an endless cyclical process of rocks forming under the sea, being uplifted and tilted, then eroding to form new strata under the sea once more: accordingly, Earth is "not just a machine but also an organised body as it has a regenerative power." He concluded that an "indefinite space of time" had been needed to produce the land that now appears from "the consolidation by means of fusion" of earlier sediments in the ocean deep; an equal space of time would be needed to destroy it. He famously stated that, in geology, "I can find no vestige of a beginning, – no prospect of an end." At that time, no meaningful conception of geological time existed: this topic will be taken up in Chapter 6. Hutton's theory became known as uniformitarianism, a polysyllabic noun coined in 1832 by the accomplished wordsmith William Whewell (1794–1866), Master of Trinity College, Cambridge.

At the beginning of the eighteenth century, travel guides and scientific instruction books for naturalists appeared. John Woodward's (1665–1728) *Brief Instructions* of 1696, and his follow-up *Brief Directions* of 1728, emphasized that field notes should list the different rocks observed at and below the surface. Observers were urged to pay particular attention to the position and the thickness of strata. Minerals and mineral veins should be noted and sites where fossils may be found described.

John Whitehurst (1713–1788), a clockmaker and engineer in the English Midlands, followed Woodward's advice and became a pioneer of local geology. He was the oldest member of the Lunar Society of Birmingham, a famous dining club founded in 1765 to stimulate convivial conversations between local natural philosophers and entrepreneurs. In this milieu he interacted with inventive individuals such as the famous potter Josiah Wedgwood FRS (1730–1795), the manufacturer Matthew Boulton (1728–1809) and the outstanding intellectual Erasmus Darwin (1731–1802), the grandfather of Charles, the future naturalist. In 1778, Whitehurst published the results of several years of geological research, much of it being fanciful speculation, but in his "Appendix on the strata of Derbyshire" we find the following gem:

> It may appear wonderful that amidst all the confusion of the strata, there is nevertheless one invariable order in the arrangement of them, and their various productions of animal, vegetable and mineral substances … By knowing the arrangement and affinities of the strata, we are enabled to investigate, with much certainty, whether coal or limestone are contained in the lower regions of the earth.[1]

⊙ THE COAL PROSPECTOR AND THE GEOLOGICAL MAP

Our next pioneer is William Smith (1769–1839), who was variously a drainage engineer, land surveyor and coal prospector (Figure 5.3). Before the late eighteenth century, English landowners (the gentry) didn't take much interest in the rocks and minerals under their feet,

FIGURE 5.3 William Smith, aged 69. Stipple and line engraving published 1837 by Ackermann & Co., London. *Source:* National Portrait Gallery, London NPG D6788. Creative Commons license

although the more enterprising among them were capitalizing on the improvement of agricultural land through drainage schemes and the introduction of machinery. From around 1760, the energy unleashed by Britain's Industrial Revolution was drawing landowners' scrutiny downwards in the search for economically useful coal deposits. New industrialists sank mine shafts in their quest for coal – the fossilized remains of ancient swamps – for which the steam engines, pumps and the iron industry had an insatiable appetite. New canals cut across the countryside to transport coal efficiently. Geology and mineralogy flourished side by side. Fieldwork advanced from tapping the surface rocks and minerals with handy hammers to the more thrilling prospect of viewing the newly exposed stratified rocks unveiled by the cutting of canals, mining operations and quarrying for building materials. The operational needs of the mining industries led to precise instructions on how exploration mineralogists should prepare their expeditions: what to pack (as little as possible), what instruments and chemicals would be essential in the field and the importance of notebooks for precise data collection in the field.

In 1791, we find Smith in Somerset, a county in England's pastoral West Country. He would stay in the area for eight years, making a profitable living by surveying mines and the proposed route of a canal to transport coal to London. His initial commission entailed a survey of Sutton Court, an extensive family estate with a mansion house dating from the fourteenth century. Smith had a head start thanks to a geological survey that had already been undertaken 70 years earlier by John Strachey (1671–1743), who had inherited his father's estate, together with its extensive coal workings. Strachey had dashed off a letter and copy of the cross-section of his survey to a friend in London, one Robert Welstead FRS, who helpfully communicated it to the Royal Society at its meeting on May 7, 1719, after which it appeared in *Transactions*, under the title "A curious description of the strata observ'd in the coal-mines of Mendip in Somersetshire." Its engraved illustration shows all of the features still in use today in a standard stratigraphic cross-section, and displayed for the first time the order and regularity of strata (Figure 5.4).[2] Smith had

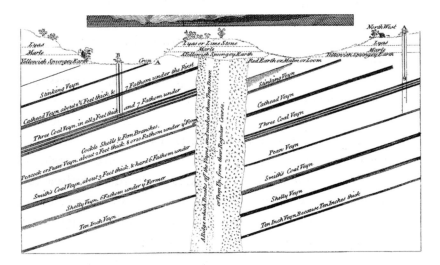

FIGURE 5.4 John Strachey's stratigraphic cross-section from the southeast to the northwest of the Upper Carboniferous Coal Measures in the Mendip Hills of Somersetshire, southwest England, published in 1717. The coal seams (veyns) dip to the southeast and are unconformably overlain by horizontal beds of Mesozoic age. A wide fault zone is portrayed beneath the central topographical ridge, which offsets the coal beds.

Source: Strachey (1717)[2]

already learned sufficient geology to be able to infer the location, thickness and dip of the land's two distinct areas of coal measures. From conversations with colliers he understood that the seams being worked on in the two separate places were actually the same body, but at different depths. Strachey's engraving, plus advice from the coal face and his own observations at the surface, enabled Smith to sketch a cross-section of the subterranean arrangement of the coal seams.

Some 70 years later, Smith viewed his predecessor's diagram as a template for his own investigations, knowing that a commercial advantage would be gained from understanding the distribution of strata. As he crawled and clambered through the underground mine workings, he amassed a reference collection of fossils, supplemented by his examination of exposures in canal, road and railway cuttings, as well as quarries and escarpments; his collection of fossils survives in the Natural History Museum, London. Smith surveyed routes for the Somerset Coal Canal in 1794, and he was on hand when excavation started in 1795. Deep surgery of Earth's strata revealed similar sections of rock and clay in two branches of the canal, and as the surveyor he knew the level of the cuttings. By the end of 1795, he was fully familiar with the local strata, and he worked on the order of the layers, starting with chalk strata at the top and descending to lime-stone below the coal seams. He interrogated the fossil record to identify the different strata in similar rock formations. Smith began to recognize that sedimentary rocks could be identified by the fossils associated with them, and that these rocks were always arranged in the same order. Although coal miners were already aware of the succession of coal seams, it was Smith who made the mental leap of decoding the fossil records in mudstones to determine the order of the strata. This was a huge breakthrough for the science of geology. It was immediately obvious to Smith that the coal occurred in association with grey mudstones. But there were also mudstone strata far below and above the coal. From the embedded fossils he could now identify the characteristic features of the mudstone associated with coal, for which demand was soaring, and thus profit from deep carbon deposits.

⊙ DEFINING THE ORDER OF THE STRATA

In 1798, Smith noticed a small circular map in the *County Agricultural Report*.[3] The map had a diameter of 40 cm and it displayed not only hills, rivers and lakes, but also rocky outcrops, including coal measures. The concept of using color to demarcate different soils and rocks at the surface inspired Smith to extend the technique to a third dimension – depth – to display the courses followed by the layers of rock below. By the early summer of the following year, Smith laid out on his desk a newly available geographical map of the Bath area. On this he painstakingly began to superimpose the geology of the area. From the lie of the land, as well as his field notes on the dip and strike of the strata, he deduced where the various strata lay. His first map showed the outcrops of just three types of rock, with the various subterranean layers rendered by Smith's hand-coloring: the oolite in deep yellow, the lias in blue and the trias in brick red. It is the oldest geological map made in Britain.

At the annual gathering of the Bath Agricultural Society, Smith's display of "A Map of Five Miles round the City of Bath" caught the eye of Benjamin Richardson (1758–1832), the vicar of a rural parish nine miles south of Bath. He invited Smith to admire his hoard of fossils and rocks. To Smith's dismay, these were organized according to the type of fossil: ammonites stacked up here, crinoids (similar to modern starfish and sea urchins) piled in drawers over there, belemnites (cephalopods) underneath and so on. This would not do: Smith immediately set to work on rearranging the amateur paleontologist's jumble according to the stratigraphic order, with the oldest at the bottom and the youngest at the top. It took him just a day. Richardson was amazed at the transformation, and even more so when the young prospector–surveyor told him of the beds to which particular fossils belonged. He outlined the principle of superposition: that the same characteristic fossils are always found in the same strata, and these strata are arranged in the same subterranean order, independent of geographical location.

Without further ado, Richardson contacted his friend Joseph Townsend (1739–1816), an imposing clergyman of towering intellect with a fascination for mines: his father Chauncy Townsend (1708–1770) had made a fortune as a merchant, which he then lost by extending his coal mines in Wales. Townsend had amassed one of the largest fossil collections in the country. He passionately believed in the Mosaic account of the Creation; this rock-solid faith inspired him to search for evidence that would vindicate Moses as a true historian. By poking around in Somerset's mines and quarries, the scriptural geologist hoped to unearth features and fossils of the global Deluge, the tangible evidence of God's handiwork. On learning from Richardson of Smith's interesting conclusions about strata, he decided to test Smith in the field: What rocks and fossils would be found on the slopes and summit of the nearby Mendip Hills? Richardson hired a horse-drawn carriage, and the three set out for Dundry Hill, a prominent landmark south of Bristol. They paused several times on the slopes: time and again the outcrop, its fossils, the lithology, and the successions were exactly as Smith had predicted.

Following that productive field excursion, Townsend arranged for the three of them to dine at his residence, 29 Great Pulteney Street, a magnificent townhouse with a honey-colored facade of Bath freestone (oolitic limestone). When Smith arrived in a downpour of rain, he was not in a sunny mood: the Somerset Coal Canal Company had summarily dismissed him, possibly because of a conflict of interest over a land deal he had made. Townsend may have hoped his dinner invitation would lift Smith's spirits, so he could pull himself together. We don't have a record of the menu, but we can safely assume that in Jane Austen's Bath, Reverend Townsend would have served a hearty repast of local fare, followed by refined discussions over port and Madeira wine.

Smith reviewed what they had seen at Dundry Hill. They concurred with his discovery of the regular order of strata. That observation validated his technique of using fossils to classify each "peculiar Stratum." As a seasoned prospector for coal, he had indicated how stratigraphy could now confer real commercial advantage. The two divines urged him to publish his scheme, both to establish his priority and to ensure that he would receive due credit and reward. Townsend had prepared for what happened next: he unrolled a large sheet of paper and had quill pen to hand. Townsend ruled a large table and looked up expectantly at Smith, who rose to his feet to dictate his ordering of 23 horizons, from chalk down to coal. His descriptions of each band of rock included its thickness and his observation of fossils that were unique to that layer, along with its situation and where the formation could readily be seen. For example, here's how Smith describes Bath freestone:

> Top-covering anomiae with calcareous cement, strombites, ammonites, nautilites, cochliae hippocephaloides, fibrous shell resembling amianth, cardia, prickly cockle, mytilites, lower stratum of coral, large scollop, nidus of the muscle with its cables.

He helpfully listed the localities for seeing the freestone: Claverton Down, Swanswick, Batheaston, Hampton and a dozen more. Finally, for layer 23, coal, we have:

> Impressions of ferns, olive, stellate plants, threnax-parviflora, or dwarf fan-palm of Jamaica.

Among these layers, the men noted something that struck them as surprising. Between strata they called Millstone and Pennant Stone, the kinds of fossils embedded in the rocks changed dramatically. In Smith's words, beneath the Millstone bed, "no fossil, shells or animal remains are found; and above it, no vegetable impressions." In the Pennant Stone layer, marine mollusk shells dominated. Today, we know that the turnover of fossils from plants to seashells is the boundary between the Carboniferous, with its abundant swamps, and the Permian, when all landmasses were collected into the super-continent Pangaea. In Bath, on that momentous evening, the trium-virate were unconsciously committing to paper the evidence for the end of one period of geologic time and the beginning of the later

Permo-Triassic period. The boundary between the two was uncon-
formable, with a hiatus between episodes of sediment deposition.
Earth therefore had a history, written in the rocks, and immensely
longer than the history the divines knew from biblical sources. The
scientific exploration of the matter below Earth's surface and its
internal workings thus began in earnest at the close of the eighteenth
century.

⊙ THE FIRST LARGE-SCALE GEOLOGICAL MAP

It would take half a century before fossils were universally accepted as
organic in origin and therefore could be considered as "documents"
recording Earth's history. Smith never used fossils to reconstruct past
geological events: first and foremost, he was a commercial mineral
prospector, searching for deep carbon in the form of coal. He had
recognized, from 1793, that in southwest England the strata lie in a
regular succession, inclined eastward, and terminating at the surface:
his homely analogy being that they "resemble, on a large scale,
the ordinary appearance of superposed slices of bread and butter."[4]
With his engineering background, he was interested in the three-
dimensional arrangement of the strata, their outcrops and their
landforms.

In June 1801, Smith issued a book prospectus for *Accurate
delineations and descriptions ... of the various strata ... of England
and Wales*. He took to traveling extensively for fieldwork while
continuing to support himself by jobbing as a surveyor, charging two
or three guineas a day to landowners hoping to amass riches from
their unseen mineral deposits (hardly any did). He clambered aboard
stagecoaches and dropped into posthouses or inns for his evening meal
and a bed. During the most intense phase of his mapping activity, he
was notching up 15,000 km a year, while carefully noting the surface
features of over 120 km^2. Between 1794 and 1821, he produced a series
of excellently printed, detailed maps of 15 counties. These had only
limited circulation, so Smith decided to publish on a smaller scale and
with less detail a geological map of England and Wales.

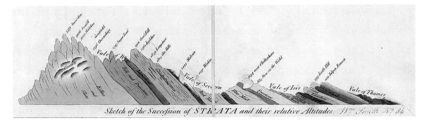

FIGURE 5.5 William Smith's "Sketch of the Succession of Strata" from Snowdonia (left) to the Thames Valley (right). This image appears as a detail on the famous 1815 map, *A Delineation of the Strata of England and Wales, with Part of Scotland*.

In August 1815 in London, the leading cartographer John Cary (1755–1835) published on 15 large sheets the first edition of Smith's gigantic geological map (Figure 5.5). This was the first attempt to represent on a large scale the geological relations of an extensive area. Without question it was a magnificent achievement, and it set the template for subsequent geological maps.[5] In an explanatory text of 50 pages, he introduced a stratigraphic terminology, and many of his names for horizons have remained to the present day. Smith pioneered the geological map as an aid to our understanding of the history and formation of Earth's geology and therefore its structure immediately below the surface.

☉ PROFESSING EARTH SCIENCE

In the nineteenth century, the emerging discipline of geology progressed remarkably. In Great Queen Street in London's West End, a small scientific dining club met at the fashionable Freemasons' Tavern. It was here that the Geological Society of London was launched by Humphry Davy (1778–1829), the English chemist who, in 1807–1808, had isolated (or discovered) the elements potassium, sodium, magnesium, calcium, strontium and barium. He suggested that the geologists should meet regularly for a dinner followed by discussions. This format would be a less ponderous alternative than the somnolent formalities of the Royal Society's meetings, at which papers were recited wearily by an official reader, but only after the completion of routine business such as nodding approval of the minutes of the previous meeting and applauding the list of presents (books, specimens and so on) recently received.[6] A small group of mineralogists had already been meeting for working breakfasts, but Davy advocated a change to an evening meeting of "a little talking Geological Club," noting that:

> The chills of November mornings are very unfavourable to ardor in the pursuit of science, and I conceive that we should all talk and think better after experiencing the effects of Roast Beef and wine rather than in preparing for tea, coffee and Buttered Buns.

Thirteen enthusiasts inaugurated the Geological Society on November 13, 1807. It was the first learned society in the world to be devoted to geology, and it quickly began to accumulate a library, as well as a reference collection of rocks, minerals and fossils. At the first meeting of its Council, the Society decided to publish important communications. The first volume of its *Transactions* came off the press in 1811. It published the text and fine illustrations of geological surveys, as well as fieldwork papers selected from those read at meetings.

The foundational principles of the Society were to stimulate the zeal of geologists, to induce them to adopt systematic nomenclature and to ascertain "what is known in [geological] science and what yet remains to be discovered" – its motto is *Quicquid sub terra est* ("Whatever is under the earth"). English and Scottish geologists soon dominated the invention of names for geologic periods, eras and epochs in a stratigraphic sequence, and their terminology is still in use today. The collection of data began straightaway with the distribution of *Geological Inquiries*, a small booklet that sought locally sourced information on deposits. Such information flowed into the Society, from members and nonmembers alike. Its first President, George Greenough (1778–1855), was so pleased with the level of support that he formed a national group of observers by enrolling the hobbyists as formal members. This is the earliest example in the Earth sciences I have been able to find of empowering a citizen network to harvest data and specimens. Membership of the Geological Society increased rapidly, hitting 341 in 1815, and then 400 three years later. King George IV graciously granted the Society a Royal Charter in 1824, and from 1828 his government provided prestige apartments in the North Wing of Somerset House. In Britain, the emerging discipline of geology thus began to coalesce into an academic field underpinned by royal patronage.

In 1808, a handful of members of the Society, led by its President, Greenough, visited William Smith's London home to view his fossil collection and regional maps. They were weighing up whether Smith was capable of making an official geological map of the entire country. They came away unconvinced by the prospector's theories on stratigraphy, nor were they impressed by his slow rate of progress as a cartographer. In any case, Smith was not of their social class – he could ill afford the 15-shilling monthly fee to attend the Society's dinners. The Society therefore turned to its President and commissioned him to produce the official geological map of England and Wales. Unlike Smith, Greenough did not get his hands dirty with fieldwork, relying on others to send him rock specimens. After 1815, he could craftily consult Smith's map, which led to allegations of plagiarism when the Society's official map finally came off the press in May 1820.

Meanwhile, in 1815, Smith fell on desperately hard times: sales of his own map were disappointing, and he had made a disastrous investment in a quarrying enterprise that failed. A deep recession at the end of the Napoleonic Wars dampened the spirit of the English landed classes for investment in coal mining. In the summer of 1819, when his cash ran out, Smith was confined for several weeks in the King's Bench Prison, Southwark, until the sale of his London townhouse was completed. At the Society, Greenough and the elite continued to doubt the usefulness of fossils for learning anything about rocks.[7] In bald terms, "Its leaders at first did not believe he [Smith] had uncovered anything of significance and then they simply stole much of it."[8] Smith was quietly forgotten. In 1824, he slipped away to Scarborough in Yorkshire, where he made a living as an itinerant surveyor.

Happily, with the passage of time, a different generation of geologists succeeded the founder members of the Geological Society. At a special meeting on January 11, 1831, the Council resolved unanimously to award their first Wollaston Medal to Mr. W. Smith:

> ... in consideration of his being a great original discoverer in English geology, & especially for his having been the first [in this country] to discover & to teach the identification of strata, & their succession, by means of their embedded fossils.[8]

The Wollaston Medal is the Society's highest award. At the February 1831 meeting, Adam Sedgwick (1785–1873) ceremonially presented Smith with the 20-guinea purse but not an actual medal, because it was still in production at the mint. And in return, William Smith gifted the Geological Society three documents: his original manuscript of the table of strata, the small circular map of Bath and a preliminary version of his great geological map. He had finally achieved professional status.

⊙ THE MAGNETIC CRUSADE: THE FIRST INTERNATIONAL GEOPHYSICAL SURVEY

The gradual growth of geological fieldwork in Britain, Europe and Russia facilitated the organization of data collection on a systematic basis across the European continent. In the early nineteenth century, the British Army, then scattered across the enormous British Empire, offered plenty of opportunities overseas to participate in campaigns that were launched to improve knowledge of Earth's magnetism by coordinating observations throughout the Empire. Since the days of Halley's two voyages, it had been widely recognized that Earth's magnetic field was continually changing in a complicated fashion that affected compass readings at sea and on land. Empires, international

FIGURE 5.6 Alexander von Humboldt. From the portrait of 1843 by Joseph Karl Stieler. *Source:* Schloss Charlottenhof, Potsdam

trade and maritime warfare all called for a more comprehensive understanding of geomagnetism and the magnetic compass.

With that in mind, the Prussian polymath, explorer and natural historian Alexander von Humboldt (1769–1859) (Figure 5.6) made many geomagnetic observations in the course of his four extensive expeditions. By the time of his final exploratory journey (1829), Humboldt had the ambition to foster international cooperation by establishing magnetic observatories. Synchronized observations began in observatories in Berlin, Freiberg and Paris in 1829. In November of that year, Humboldt was in St. Petersburg, where he persuaded the authorities to set up a network of magnetic observatories throughout the Russian Empire. By the mid-1830s, a chain of 11 observatories spanned the Empire, from St. Petersburg in the west to Sitka, the capital of Russian Alaska, in the east.

Humboldt encouraged the great mathematician and physicist Carl Friedrich Gauss (1777–1855) to establish the Göttingen Magnetic Society, which flourished from 1836 to 1941. Gauss and the German physicist Wilhelm Eduard Weber (1804–1891) worked together on building a network of stations for globally synchronized measurements of the intensity of the geomagnetic field. Data were acquired at up to 53 observatories on agreed dates, with readings taken every 5 minutes. It was the earliest example of an international scientific collaboration. On August 18, 1840, an intense "geomagnetic storm" was registered; Humboldt speculated that these strong changes in the magnetic field could not originate from Earth's interior and must be caused by solar activity.

When the association became moribund in 1841, the task of collating geomagnetic observations was assumed by the Royal Society. The background to this was that by the late 1830s the Royal Society and the Royal Navy were each devoting so much energy to solving the problem of magnetic variation that it had become a "British" rather than a "continental" science.[9] The key person who engineered this was Edward Sabine (1788–1883), an Anglo-Irish astronomer, geophysicist, army officer in the Royal Artillery and President of the Royal Society (1861–1871). He had tasted action in the war of 1812 against the USA when he commanded the batteries at the Siege of Fort Erie, where the British took heavy casualties. On his return to England in 1816, he devoted the rest of his long life to geophysics and ornithology. Sabine displayed a remarkable ability for mingling with men of influence in scientific and military circles, and his rapid acceptance into the scientific establishment led to his election to the Royal Society in 1818. Next, in 1818–1820, he gained considerable experience of geomagnetic research as a scientific officer on two expeditions financed by the Admiralty to seek the Northwest Passage. Two decades later, we meet him as the driving force behind a campaign to organize a Magnetic Crusade as a global enterprise.

The support of the wider scientific community in London for the Magnetic Crusade, and particularly that of the astronomer Sir John Herschel (1792–1871), who had strong political connections, was critically important for exerting political pressure on the British Government to secure funding for science on a magnitude never seen before in Britain. Sabine and Herschel were fanatical in their commitment to ensuring that the funding would deliver what was promised in terms of raising the international profile of British science.

In spring 1839, the government of the young Queen Victoria (then aged 20) acted to establish magnetic observatories throughout

Victoria's colonies and dominions. It turned out to be very challenging to set up the colonial observatories and to persuade foreign observatories to participate. An international conference in Cambridge in 1845 marked the high point of the Magnetic Crusade: funding was then agreed for a final three-year campaign. In 1848, at the conclusion of the extension, Britain had a network of geophysical observatories that ringed the world. Magnetic data continued to flow in from Austria, Belgium, Russia, Prussia, Spain and Sweden.

In 1851, Sabine's analysis of the data from the colonial observatories revealed a correlation between the 11-year sunspot cycle and the frequency of magnetic storms. This was the first direct evidence that major disturbances of Earth's geomagnetic field are caused by outbursts of activity on the Sun. Later that year, Sabine discovered that the daily variations in magnetic intensity are due to two superimposed variations: one from within Earth and the other external to it. Neither of these discoveries could have been made from observations at a single station: they required a global network. Although the scientific achievements were modest in terms of discoveries made, the impact of the campaign was enormous it terms of demonstrating the potential of global scientific observation. This is the earliest example we have of international cooperation in geophysical discovery that extended over a decade in time, that encompassed a worldwide network of participants and that organized the large-scale collection of data and their analysis.

REFERENCES

1. Whitehurst, J. *An Inquiry into the Original State and Formation of the Earth: Deduced from Facts and the Laws of Nature* (W. Bent, 1786).
2. Strachey, J. A Curious Description of the Strata Observ'd in the Coal-Mines of Mendip in Somersetshire; Being a Letter of John Strachey Esq; To Dr. Robert Welsted, MD and RS Soc. and by Him Communicated to the Society. *Philosophical Transactions (1683–1775)* **30**, 968–973 (1717).
3. Winchester, S. *The Map That Changed the World: William Smith and the Birth of Modern Geology* (HarperCollins, 2001).

4. Phillips, J. *Memoirs of William Smith, LL.D.* (John Murray, 1844).

5. Zittel, K. A. *History of Geology and Palaeontology to the End of Nineteenth Century* (Walter Scott, 1901).

6. Hall, M. B. *All Scientists Now* (Cambridge University Press, 1984).

7. Laudan, R. Ideas and organizations in British geology: a case study in institutional history. *Isis* **68**, 527–538 (1977).

8. Torrens, H. S. Timeless order: William Smith (1769–1839) and the search for raw materials 1800–1820. *Geological Society, London, Special Publications* **190**, 61–83 (2001).

9. Cawood, J. The magnetic crusade: science and politics in early Victorian Britain. *Isis* **70**, 493–518 (1979).

6   Thousands, Millions or Billions
The Question of Timing

⊙ READING NATURE'S ARCHIVES

In this chapter, I assess the connections between the studies of deep carbon as the carbonate component of fossils and the role of deep time for a reconstruction of Earth history. I leave the history of the absolute calibration of deep time for the following chapter. William Smith's standout contribution to the practice of geology in England was his achievement in tracing the courses of the strata. He introduced the principle of faunal succession in his book *Strata – Identified by Organized Fossils*, a short book with colored plates published in 1816. That publication turned his private cabinet of curiosities into a public resource of the characteristic fossils in rock formations.

In England, the younger members of the Geological Society, notably Charles Lyell, Adam Sedgwick and William Buckland (1784–1856), laid rather greater emphasis than Smith had on the use of fossils to determine the succession of strata, which would become a step along the road to a geological timescale.[1,2] In Saxony, Abraham Gottlob Werner (1749–1817) (Figure 6.1), a geologist who taught at the Freiburg Mining Academy, introduced the theory of Neptunism to account for the layering of rocks in the crust. He proposed that Earth's rocks formed rapidly by catastrophic precipitation of a global ocean. The opposing view was that all rocks came from volcanic environments. Today, Werner is remembered for writing the first textbook on mineralogy since Agricola. This introduced a practical identification scheme that could be used in the field by assigning relative values to a number of properties such as hardness. Werner's classification system placed emphasis on chemical composition rather than crystallographic form.

FIGURE 6.1 Engraved portrait (1848) of Abraham Gottlob Werner, one of the founders of modern mineralogy.
Source: Wikimedia Commons. Creative Commons license

Simultaneously with William Smith's activities, Georges Cuvier was busy in France making his mark on vertebrate paleontology and comparative anatomy. Cuvier's interest was in an established tradition of research on fossilized quadrupeds. Initially, he conducted indoor research on specimens collected by others: those already in the Muséum National d'Histoire Naturelle, as well as numerous samples and drawings he received after making an appeal in 1800. Early on, Cuvier regarded fossil bones as nature's "archives" from which it should possible to read the history of Earth. A year after arriving in Paris from Switzerland to take up a junior research position at the Muséum, he had to deliver an inaugural public lecture following his election to the Institut National des Sciences et des Arts. The topic was fossil elephants, perhaps because the Muséum had just received 150 crates of specimens that Napoleon's army had looted from The Netherlands. Two elephant skulls in the booty enabled Cuvier to prove conclusively that Asian and African elephants are two distinct species. Thinking big, he picked around elephantine teeth and bones in the Muséum's rich collections. From these small fragments he made gigantic discoveries: two new species of megafauna. Both were huge mammals that had long since vanished from the face of Earth: one the mammoth and the other the mastodon (a name he later suggested).

FIGURE 6.2 Engraved portrait
of Georges Cuvier (1826), being
the frontispiece to his *Discours
sur les révolutions de la surface
du globe*.
Source: The Linda Hall Library of
Science, Engineering &
Technology, Kansas City, Missouri

Paleontology was now at a defining moment: Cuvier found overwhelming anatomical evidence to convince most naturalists that extinction was a genuine phenomenon (Figure 6.2). Hooke had speculated that many fossils represented "Species of Creatures in former Ages, of which we can find none at present." However, it took Cuvier's discoveries to terminate the long debate about whether or not fossils were remains of living species yet to be discovered. For him, fossils were compelling evidence of an earlier world, beyond the reach of the historical and biblical narratives, which had been "destroyed by some kind of catastrophe."[3,4] Cuvier realized that he must stop relying on museum specimens curated by others and get out *en plein aire* to investigate fossils in their lithological settings – in context.

From 1804, Cuvier collaborated on fieldwork with Alexandre Brongniart (1770–1847), a zoologist, chemist and mineralogist who was director of state porcelain manufactory at Sèvres near Paris. He was keen to find new sources of raw materials for the ceramics industry. The two of them worked together on the geology of the Paris basin. Brongniart did most of the boots and hammer stuff, while Cuvier vicariously joined in the fun from time to time. This was the only substantial fieldwork he ever undertook; it was a refreshing break from his vast amount of desk work as secretary of the National Institute and his huge program of writing memoirs on the anatomical classification of animals. Like Smith, Cuvier and Brongniart selected characteristic fossils and applied the principle of succession to understand their positions in the geological column. In 1808, when they presented their classic paper to the Institut, giving a biostratigraphical interpretation of the geology of the Paris basin, they took the opportunity to show a mineralogical map that picked out in color particular formations.[5] Later, upon reading a translation of their paper, John Farey (1766–1826), an English mineralogist, quipped that the French wanted to steal the credit that was due to Smith.[6] In reality, Smith's attention was focused on the *presence* of particular fossils as a proxy for the relative age of rocks. Cuvier's primary interest was the classification and *comparative anatomy* of fossils.

Cuvier and Brongniart raised the study of fossils *in situ* to a level of great significance for constructing the history of Earth. They interpreted the fossil record of the Paris basin as a succession of long periods when chalk and limestone were deposited under the sea, together with tranquil intervals of freshwater inundation that led to gypsum formations. Their fascinating story of a complex geohistory for the Paris region became a template for naturalists in other regions. Structural sequences of formations and their fossils could be presented as temporal narratives of distinct environmental periods, separated by occasional changes.

⊙ DEEP TIME AND NATURE'S DEEP HISTORY

A major problem in reading the stratigraphic column as an archival repository of Earth history was the absence of an absolute timescale. Everything was relative, which had made for half a century of lively debates. Was Earth shaped by sudden events on a gigantic scale of violence – the Deluge, earthquakes and volcanoes – or was it the product of gentler geological processes – particularly erosion and deposition – operating over an immense span of time? In the eighteenth century, enlightened natural philosophers began to realize that Earth had a history of its own, a history that did not align with any of the ancient sources. This *deep* history would have to be learned through examination of the natural world: its rocks and minerals; the mountains, volcanoes, oceans, atmosphere and continents; and the plants and animals. To reconstruct Earth's deep history with confidence would require challenging fieldwork, careful observation, precise recording of data and, above all, rigorous application of the scientific method. In short, it needed a new way of doing science, a clean break with tradition and classical philosophy.

Since antiquity, natural philosophers had considered the cosmic aspects of nature as amenable to informed observation and quantitative inquiry, most notably the motions of the planets. As early as 4000 years ago, clerks in the temples of Mesopotamia recorded in cuneiform script the changing aspects of the heavens. In the early sixteenth century, Copernicus spent four decades working on *De Revolutionibus*. Later, Galileo, Johannes Kepler (1571–1630) and Newton developed geometric approaches to planetary dynamics that were firmly anchored to the *laws of physics*. However, natural philosophers seeking the truths of Earth's history could never find so-called *laws of nature* that could be encapsulated in neat algebraic expressions. The history of Earth could only be pieced together by studying the evidence that has survived to tell us *facts* about the past. This was a critical moment in the history of Earth science: the realization that the rival theories of Earth's physical history could not be resolved by philosophical assertion alone. As we shall now see, the key to the mystery required a reliable value for the age of Earth.

☉ USSHER AND THE AGE OF EARTH

The history of anything needs a timeline. The passage of time can be represented in many ways: calendar dates, for sure, and in historical documents the names of incumbents such as pharaohs, popes or presidents have long served this purpose. In 1650, Archbishop James Ussher famously arrived at the date of 4004 BCE for the creation of Adam. Today, it is commonplace for science writers to ridicule Ussher as naive and superficial, but until the middle of the nineteenth century his biblical chronology influenced English geologists who were clerics of the Church of England. It was an interesting example of the gulf that then separated science and religion. Ussher was an outstanding classicist, hugely well read in the works of the ancient world.[7] In the context of his own times, he was an academic in the humanities working in an honorable field: there are 12,000 footnotes detailing the sources he used for his timeline of Creation. He backtracked the *historical* records as far as he was able in order to clarify the history of the birth of Jesus Christ. He showed no interest in Earth history and made no claims about the physical age of the universe. From 1701, the *Authorized King James Bible* of the Church of England included a historical marginal note in red ink of "4004 BC" for the Creation of Heaven and Earth. That's how Ussher's date became publicized and widely accepted. Hence the curious interest among geologists during the seventeenth and eighteenth centuries in correlating Old Testament "events" such as the Deluge with the record on the ground and in the rocks.

⊙ "DRAWING LARGE CHEQUES UPON THE BANK OF TIME"

By the middle of the 1800s, stratigraphic geologists had long since ceased to argue about the restless Earth. They had discarded the hypotheses of the previous generations: Werner who suggested that rocks had precipitated from a universal ocean that receded with the subsequent exposure of land; likewise, the opinions of Smith and of Cuvier, who, along with Buckland at Oxford, had a liking for catastrophism with its violent raising of mountain chains and cataclysmic floods. James Hutton and Lyell (born in the same year as Hutton's death) were cast as survivors of the great debates.

In 1788, Hutton had played a big part in introducing the scientific method to geology through a paper that he later expanded to a two-volume work.[8] This epochal book led to his being dubbed by myth-makers as the "father of modern geology." He proposed a new philosophy, to be informed by careful observation of actual circumstances and then accounted for by the long-term action of natural forces: wind, rain, tides, sedimentation, volcanic upthrust and Earth's internal heat. In his new dynamic geology, the action of agencies that had molded Earth in the past had continued into the present, undiminished in intensity. Growth and decay on Earth were in equilibrium:

> It has a state of growth and augmentation; it has another state, which is that of diminution and decay. This world is thus destroyed in one part, but it is renewed in another; and the operations by which this world is thus constantly renewed, are as evident to the scientific eye, as are those in which it is necessarily destroyed.[9]

This new philosophy of uniformitarianism denied the short timescale of the convulsive theories. Hutton developed this slow-motion, steady-state model of nature's course in the company of close friends, who themselves were intellectual giants. Together with the moral philosopher Adam Smith (1723–1790) and the chemist Joseph Black (1728–1799), Hutton was a founding member the Oyster Club, a convivial forum for Edinburgh intellectuals who met weekly in the dark taverns of the Old Town. They slipped down oysters accompanied by copious quantities of local ale, Scotch whisky and London gin. Hutton declined the spirits, but nevertheless found the discussions "informal and amusing, despite their great learning." Visiting participants at the Oyster Club included Benjamin Franklin (1706–1790), whom Hutton may have met in 1771, when Franklin stayed with the renowned philosopher David Hume (1711–1776). In his own time, Hume came under fire for his irreligious views: he was insulted as "an Atheist" and "Great Infidel." Hutton's philosophy clearly drew on Hume's total rejection of supernatural events. He also respected the stance of Hume and Adam Smith on knowledge and empiricism. Hutton, like Adam Smith, was a Newtonian: explanations of how nature works must be based on observation leading to accounts that are systematic and free of gaps.

For Hutton, deep time was never an issue. It is true that some churchmen reacted to his theory with righteous horror, but contrary to the modern myths of repression and persecution, the Church authorities in England did not criticize those who took a very long timescale for granted: they merely drew the line at an eternal and uncreated universe. Hutton's ideas did not readily percolate through the emerging communities of scientists, and that was largely his own fault. He penned a peculiar, intricate and confusing style of writing, with one of his volumes partially written in French. Perhaps for that reason it took almost 50 years for his views to attain general acceptance through the hands of other writers. His visionary prospect on deep time and long-duration Earth history finally found wide and sympathetic readership through Lyell's *Principles of Geology* of 1830–1833. This book promoted geology as a true science, with no appeal to the supernatural or speculation.[10] But Lyell, too, needed deep time.

The older generation resisted deep time for longer, but not Sedgwick at Cambridge, who very publicly switched to uniformitarianism in 1831. In 1818, the year of his ordination in Anglican orders, Sedgwick had been elected to the Woodwardian Professorship of Geology (paleontology, to be precise), with responsibility for the huge collections of fossils at the University of Cambridge. His electors included the Archbishop of Canterbury and the Bishop of Ely. As an ordained clerk, Sedgwick began his career lecturing on a history of Earth that accorded with *Genesis*, but with a tweak to the narrative, in which the biblical "days of Creation" were recast as the turning points in a divine drama of indeterminate but finite length. Charles Darwin, an undergraduate at Christ's College, attended Sedgwick's lectures on geological excursions early in 1831. Sedgwick's charismatic style reinforced the young scholar's

FIGURE 6.3 The Reverend Adam Sedgwick MA FRS, Woodwardian Professor of Geology. Mezzotint published in 1833 by Molteno & Graves, London, after the portrait by Thomas Phillips in the Department of Earth Sciences, University of Cambridge. National Portrait Gallery, London. NPG D5929.

enthusiasm for natural history, a subject to be conducted outdoors as an exercise in understanding God's creation. The reverend professor generously chose Charles as his assistant for summer fieldwork in north Wales, where he gave his recent graduate a crash course in field geology. This was to be a life-changing event for Darwin. On July 9, he exclaimed in a letter to his older cousin, William Darwin Fox (1805–1880): "What a capital hand is Sedgewick [*sic*] for drawing large cheques upon the Bank of Time!"

Lyell's strong antediluvian stance had led to Sedgwick's acceptance that rocks that formed from waterborne sediments were of various dates and not the result of a one-off global catastrophe (Figure 6.3). In his 1831 valedictory address given to the Geological Society at the conclusion of his presidency, Sedgwick proclaimed:

> We have a series of proofs the most emphatic and convincing, – that the existing order of nature is not the last of an uninterrupted succession of mere physical events derived from laws now in daily operation: but on the contrary, that the approach to the present system of things has been gradual, and that there has been a progressive development of organic structure subservient to the purposes of life.[11]

With that pronouncement from their outgoing president, British geologists could plant their boots on strong and stable ground to look back to a remote past of indefinite length. Sedgwick advised:

> We cannot take one step in geology without drawing upon the fathomless stores of by-gone time.[12]

The liberal-minded Church of England did not perceive any disharmony between religion and an inductive science based on observation of the natural world, thus quietly dismissing Ussher's chronology. By the middle of the nineteenth century, the guesstimates of the professionals still varied widely: the temporal calculus had changed from thousands of years to unspecified millions in no time at all. In a major change of exegesis, the first three words of the bible, "In the beginning . . ." were now understood as an indeterminate lapse of time before the appearance of *human life* on Earth. Another half-century would pass before surprises in physics provided new techniques to measure the age of Earth, and the story of that is given in the next chapter.

⊙ ANTOINE-LAURENT DE LAVOISIER: COMBUSTION AND RESPIRATION

In France, by the end of the eighteenth century, the quest to understand how living organisms interact with atmospheric gases came to dominate the research frontier of carbon science. The themes of coal, life and the carbon cycle were central to the research of Antoine-Laurent de Lavoisier (1743–1794), who was born into a family of wealthy merchants in Paris and later inherited a large fortune on the death of his mother. At the age of 18, he enrolled in the Collège des Quatre-Nations' law school at the University of Paris (Figure 6.4), where he busied himself with the natural sciences on the side. When he was granted his license to practice the law in 1763, he had already decided to spend his life in science rather than the law courts. One year later, he read his first paper, on the nature of the mineral gypsum ($CaSO_4 \cdot 2H_2O$), at the Académie des Sciences. He was elected an academician in 1769 aged just 26 and on the threshold of a

FIGURE 6.4 View (c. 1680) of the newly completed Collège des Quatre-Nations. Engraving by Adam Pérelle (1640–1695). A masterpiece of French neoclassical architecture, this Palais de l'Institut de France has been the seat of five French Academies since 1805.

glittering career as a chemist. In Lavoisier's student days, chemistry had barely moved on from the Late Middle Ages. The direction of chemistry was muddled by outmoded names and symbols beloved by alchemists. A major barrier to progress was the phlogiston theory of combustion, which held that in the reaction we know as combustion (oxidation) and respiration (breathing) a fire-like element is released: phlogiston (from the Greek for "burning up"). Another barrier was the belief, inherited from alchemists, that matter was transmutable. The sixth element received the name *carbone* in France in 1787 when Lavoisier, together with three highly distinguished cofounders, recommended the Académie Française to adopt a new method of chemical nomenclature that discarded the vocabulary of the classicists and alchemists. They provided names for 55 substances regarded as elemental (19 organic acids, 17 metals, light, caloric and so on) as well as the noun terminations "ic" and "ous" for acids (nit*ric* or nitr*ous* acid), together with "ate" and "ite" for their salts (copper sulf*ate* or sulf*ite*).

In his notebook, Lavoisier wrote that on September 10, 1772, he had carried out an experiment on the combustion of phosphorus.

He then wrote a sealed note to the Secretary of the Academy informing them that when phosphorus and sulfur burned, there was a gain in mass that he attributed to them combining with "atmospheric air." A few months later, he wrote that the "respiration of animals [and] combustion" were processes that fixed atmospheric air (CO_2). Next, he started a long series of experiments on combustion and respiration. In August 1774, Joseph Priestley (1733–1804) met Lavoisier and his circle. The visitor engaged them with the news that by heating red mercuric oxide he had driven off a gas that was very vigorous in supporting combustion and so pure that it might be useful for medical purposes. Priestley thought it might be the gas we know as nitrous oxide. On hearing this, Lavoisier thought otherwise, but remained silent: more likely it was the missing link he was seeking to explain the chemical reaction of combustion. He would later give the name oxygen to the gas.In the 1780s, Lavoisier and his assistants undertook a series of experiments to learn more about respiration. The most celebrated of these took place in 1784. In association with his younger colleague, Pierre-Simon Laplace, Lavoisier designed a calorimeter to measure the amount of heat released during combustion and respiration. This calorimeter had a specific amount of ice for the heat to melt and a container to collect the fixed air (CO_2). Two sources of heat were tested: a piece of charcoal (*charbone*) and a guinea pig. From measuring the heat and the quantity of fixed air, Lavoisier concluded that respiration is a slow version of combustion. The next question was: How can the quantity of oxygen used in respiration be measured? For that, he needed a human "guinea pig." His young assistant, the chemist Armand Séguin (1764–1835), breathed oxygen through a face mask while Lavoisier observed the increase in breathing rate and pulse (Figure 6.5).[13] From this experiment, Lavoisier and Séguin became convinced that respiration is slow-burn combustion, a chemical reaction that required an input of fixed air (oxygen) and an output of carbonic acid gas (carbon dioxide).

Their observations had a direct application concerning the comfort of indoor public places. Lavoisier wrote in 1785 on "alteration of

FIGURE 6.5 Lavoisier's respiration experiment on his young assistant.
This drawing by Marie Anne Pierrette Lavoisier depicts her husband in his
laboratory.
Source: Wellcome Foundation

respired air," which was based on his observations at the Palais des
Tuileries after a theatrical performance: oxygen in the upper part of
the theater became diminished and carbon dioxide accumulated. He
calculated that all of the oxygen would have been exhausted after four
and a half hours in the absence of fresh air from outside. That observation may have led him to an imaginative speculation: a guinea pig
in a bell jar perishes after an hour, humans would suffocate in a few
hours in an airtight room ... yet, after untold millions of years, plants
and animals have not yet exhausted the vitality of the atmosphere.
How come? In 1792, he looked into plant nutrition and growth, a topic
that had become trendy throughout Europe (although photosynthesis
had not yet been discovered). In a report on vegetation, he conjectured
that plants probably had "the means nature uses to maintain the
respirability of air" on a global scale.

In his final paper on coal, written a year before he was guillo-
tined on May 8, 1794, after a show trial, Lavoisier applied quantitative
reasoning to the global transfer of carbon by natural processes:

> We can conceive what immense quantity of carbon is sequestered in the womb of the Earth, since marbles, limestones and calcareous earths contain about 3/10th and sometimes 1/3 of their weight in fixed air, and this latter is composed for 28/100th of its weight of carbon; then, it is easy to conclude that the calcareous rocks contain 8 to 9 pounds of carbon by quintal.

In a geochemical aside, he remarked that there could be no doubt that elemental carbon "is part in a number of combinations in the three kingdoms – minerals, plants and animals," although he did not outline the changes in the form carbon must take to pass from the mineral (coal) to the planet and animal kingdoms.[14]

⊙ IS COAL'S ORIGIN BIOGENIC?

By 1780, a biogenic origin for coal was widely accepted in Europe. The thinking in Britain lagged behind that of the continental chemists until 1795, when Hutton made it the touchstone of his theory of Earth. Coal intrigued Hutton because it was a mineral that could not be formed through the action of water. He presented coal as a mineral that could only be produced through the action of heat on oily and resinous substances in the strata. He suggested, "The production of coal from vegetable bodies ... is made by heat, and by no other means, so far as we know."[9] Back in Paris, Alexandre Brongniart's son Adolphe-Théodore (1801–1876) followed Lavoisier's lead by urging that "geologists should also make use of botany." And, accepting his own advice, he used stratigraphy to chart the total history of plant life. The oldest set of fossils that we can easily recognize as plants are to be found in the Coal formation (Carboniferous): flowerless plants (cryptogams), many of them tall trees. Ferns and horsetails are their closest living relatives today. The younger Secondary formations are where cycads and conifers (gymnosperms) made their appearance. Finally, the flowering plants (angiosperms) become abundant in the Tertiary formations (Cenozoic era). This orderly progression of vegetable species, tracked across successive periods, mirrored what had already been found in the animal kingdom by Cuvier and several others.

☉ PARIS: A WALK IN THE PARK

During the half-century from 1780 to 1830, the savants concentrated on paleontology. The changes in the appearances of species over time greatly fascinated them. Statues of two of Lavoisier's near contemporaries are in the Jardin des Plantes, Paris. A visitor at the west entrance immediately encounters an impressive nobleman seated on a grand chair, looking imperiously into the distance. In his left hand he holds a dove, and his right hand is about to tickle the head of a lion nestling peacefully beneath his throne. This is Georges-Louis Leclerc (1707–1788), known as the Comte de Buffon after Louis XV raised him to nobility in 1772. In 1739, he was appointed director of the Jardin du Roi, a position that he held with great distinction until the end of his life. During his tenure, he transformed the Jardin into a major research center focused on natural history. Buffon brought into the realm of science the idea of life adapting to particular environments. By thinking about Earth and the presence of its life in ways that few had done before, he was responsible for promoting a substantial timescale for the history of Earth. His own estimate of the age of Earth – 70,000 years – then seemed immense. Nevertheless, Buffon influenced the next two generations of naturalists in Europe, one of whom was Jean-Baptiste Lamarck (1744–1829).

In the northeast entrance to the Jardin you can also admire a monumental bronze statue of Lamarck, erected in 1908, resting on a massive stone plinth: his posture is that of a philosopher deep in thought, gazing into the distance across the Seine. This French naturalist was consecutively professor of botany from 1788, and then of zoology from 1793, when the republican French National Assembly reconfigured the research institution associated with the Jardin as the Musée d'Histoire Naturelle. His plinth informs us of the publication of his natural history of invertebrates in seven volumes between 1815 and 1822. Today, he is remembered for his theory of the

FIGURE 6.6 Statue in the Jardin des Plantes, Paris, sculpted in bronze by the artist Jean Carlus to mark the centennial of Buffon's death.
Source: Simon Mitton

inheritance of acquired characteristics. Lamarck's philosophy considered that the environment is the force driving change in the animal kingdom and that life is structured in an orderly manner. In an imaginative step, he proposed that there exists a force (or tendency) for animals to adapt to their environments. He next set out an organizational framework for his proposal: a progressive vertical chain, ascending from the lowest (simple) forms to the highest (complex) forms of life. Lamarck's philosophy became controversial, and it was in conflict with Cuvier's that rejected evolution.[15]

⊙ CHARLES DARWIN: GEOLOGIST AND NATURAL HISTORIAN

In the middle of the nineteenth century, a breathtaking revolution in biology amplified the geologists' demands for large drafts of time: Darwin's theory of evolution by the natural selection of species. Let's return to Darwin's summer experience of 1831, when kindly Sedgwick introduced him to hands-on geology in the field. That summer job – an internship – thrilled Darwin so much that he made a decisive break from continuing to study the classics and theology: he no longer wanted to become a churchman, as his wealthy father had planned. In December 1831, 22-year-old Darwin joined the Royal Navy's HMS *Beagle* as the ship's naturalist and as a companion of the captain, Robert FitzRoy (1805–1865), who was four years his senior. FitzRoy's exceptional scientific ability inspired the Admiralty to place him in command of the *Beagle*. Launched in 1820 as a 10-gun, 2-masted sloop-of-war, she was refitted for coastal hydrographic surveys in 1825. On her first voyage as a survey vessel in 1828, her master, Captain Pringle Stokes, became so overwhelmed by loneliness and the stress of carrying out a difficult survey in the desolate waters of Tierra del Fuego that he shot himself in a fit of despair. This suicide precipitated FitzRoy's rapid promotion from flag lieutenant to captain. In preparation for his second voyage to take soundings along the coasts of South America he requested permission to have "some well-educated scientific person" on the manifest. FitzRoy wanted to recruit a geologist because privately he hoped to discover minerals of commercial value and to validate the story of the Great Deluge in the *Book of Genesis*. Inquiries of professors at Cambridge produced "a young man of promising ability, extremely fond of geology and indeed all branches of natural history": Charles Darwin.[16]

FIGURE 6.7 The survey ship HMS *Beagle*, sketched by Darwin's shipmate and first officer, John Clements Wickham.

FitzRoy was delighted to welcome Darwin. When he joined the *Beagle*'s company of some 80 officers, men and boys, FitzRoy presented him with the first volume of Lyell's *Principles of Geology*, hot off the press. After his first 20 days at sea, Darwin put Lyell's textbook to good use on a month-long visit to the Cape Verde Islands. On the largest island, Santiago (now St Jago), he found limestone cliffs embedded in recent lava flows, with fossils similar to the shells in the sea nearby. Here was evidence of the principles formulated by Lyell, and thus Darwin became convinced that the surface of Earth changes only slowly and gradually over time. By considering the uplift and subsidence of oceanic islands, he deduced that the footprint of oceanic islands can change gradually over time.[17] In the five years of the voyage of the *Beagle*, Darwin wrote 1383 pages of notes about Earth science – almost four times as much as his 368 pages of notes on natural history. The thousand sheets of geological observations empowered Darwin to grasp two fundamentals that he would follow for his scientific theory: the importance of deep time and the presence of slow, but perpetual, changes of Earth and its life-forms.

⊙ EVOLUTION BY NATURAL SELECTION

By the time *Beagle* returned, Darwin was already a celebrity. His wealthy father organized his investments to enable his talented son to become a well-funded independent scientist. Darwin then gave top priority to organizing the collections of and seeking experts from London institutions. For example, he presented his mammal and bird specimens to the Zoological Society of London on January 4, 1837. A little later, ornithologist John Gould (1804–1881) reported to the Society on his examination of the mixed bag of bird skin specimens from the Galápagos Islands: what they included was "a most singular group of finches, related to each other in the structure of their beaks, short tails, form of body, and plumage." Gould also told Darwin that the mockingbirds from the Galápagos were representatives of distinct species, each inhabiting its own island. Darwin was deeply impressed by the manner in which the birds differed slightly from island to island, given that "none of these islands appear to be very ancient in a geological sense." By July 1837, Darwin had commenced a notebook on the origin of species, sketching out his idea of a "tree of life" encompassing all species. He worked on it incessantly for the next 20 years.

The theory of evolution Darwin presented in 1859 in *On the Origin of Species* built on the work of his many predecessors. In the sixth edition (1872), he acknowledged no fewer than 34 authors who had considered the modification of species over time. Steno, Smith, Cuvier, Hutton and Lyell had amply shown that different groups of animals characterized successive epochs of Earth history: the trilobites, ammonites, dinosaurs and mammals. The absence of trilobites – the greatest survivors in Earth's history – and woolly mammoths in the modern age signaled that a steady trickle of extinctions had occurred, punctuated by the occasional catastrophic loss of species. This observation undermined a belief, stretching as far back as Aristotle, that every species has unalterable characteristics. Nevertheless, Darwin picked up Aristotle's argument that things in nature that *appeared* to have been designed for a purpose, such as teeth for masticating food, were *accidental*. That's because they had been "constituted by an internal spontaneity," rather than an intrinsic fitness for purpose. Here's Darwin again, perceptively pointing out: "We here see the principle of natural selection shadowed forth," but, he continues, Aristotle did not have the knowledge to comprehend this.[18]

Darwin recognized that his readers would not find his theory credible unless he could convince them that Earth's surface is very old. The gradual evolution of species needed a great deal of time, but how much time? Darwin attempted to estimate the passage of time from current rates of deposition and erosion. Data were becoming available from the study of sediments carried by rivers. How long had it taken these grindingly slow processes to sculpt contemporary geological features? He employed this technique to calculate the age of the Weald, a sandstone formation in southern England.

The exposed sandstone was once covered by a thick layer of chalk still visible in nearby formations such as the White Cliffs of Dover. Based on estimates of the rate of erosion, Darwin figured that at least 300 million years had expired during the removal of the chalk covering the Weald. This was three times greater than even the most generous notions of the age of Earth. Historically, it doesn't matter that Darwin's age was an underestimate: his achievement was to derive that age by adopting Lyell's concept of gradual change and then calculating how long it would have taken erosion to change the landscape from the original to the present state. This use of a natural clock to calibrate the geological timescale was a first.

Thus emboldened by his own observations and calculations, Darwin emphatically opens his famous book with the following instruction, which appeared in every edition of *On the Origin of Species*:

> HE WHO CAN READ Sir Charles Lyell's grand work on the Principles of Geology, which the future historian will recognize as having produced a revolution in natural science, yet does not admit how incomprehensibly vast have been the past periods of time, may at once close this volume.

Perhaps that was a dig at the conservative churchmen. Darwin is saying that Earth's history *and the history of life on Earth* are "incomprehensibly vast." FitzRoy, who continued to be a deeply religious man, unfortunately became depressed by the thesis set out in Darwin's book and by the growing support for evolution. On Sunday April 30, 1865, he took his own life.

Darwin's theory of evolution by natural selection emphasized the question: How old is the world? We have come a long way to answering that. Not until the next century would an absolute calibration become available. This reckoning would be achieved through developments in fundamental physics that led to marvelous inventions in geochronological techniques that continue to improve our knowledge of the history of the dynamic depths of our planet. This takes us to the next chapter on the history of understanding the deep Earth, where we begin with the entry of physicists from the mid-nineteenth century, followed by the contributions of chemists in the early twentieth century.

REFERENCES

1. Laudan, R. Stratigraphy without palaeontology. *Centaurus* **20**, 210–226 (1976).
2. Rudwick, M. J. *Earth's Deep History: How It Was Discovered and Why It Matters* (University of Chicago Press, 2014).
3. Cuvier, G. Mémoire sur les espèces d'Eléphans tant vivantes que fossiles. *Magasin encyclopédique* **2**, 444 (1796).
4. Cuvier, G. Mémoire sur les espèces d'éléphans vivantes et fossiles. *Mémoire de l'Academie des Sciences Paris* **2**, 1–22 (1799).
5. Rudwick, M. J. Cuvier and Brongniart, William Smith, and the reconstruction of geohistory. *Earth Sciences History* **15**, 25–36 (1996).
6. Farey, J. Geological remarks and queries on Messrs Brogniart's memoir on the mineral geography of the environs of Paris. *Philosophical Magazine* **35**, 113–139 (1810).
7. Gould, S. J. Fall in the house of Ussher. *Natural History* **100**, 12–21 (1991).
8. Hutton, J. Theory of the Earth; or an Investigation of the Laws observable in the Composition, Dissolution, and Restoration of Land upon the Globe. *Earth and Environmental Science Transactions of The Royal Society of Edinburgh* **1**, 209–304 (1788).
9. Hutton, J. *Theory of the Earth: With Proofs and Illustrations.* **1** (Library of Alexandria, 1795).
10. Rudwick, M. J. S. The strategy of Lyell's *Principles of Geology. Isis* **61**, 5–33 (1970).
11. Sedgwick, A. Address to the Geological Society, delivered on the evening of the 18th of February, 1831. *Proceedings of the Geological Society of London* **1**, 305–306 (1831).
12. Sedgwick, A. Appendix: The second of three letters to William Wordsworth on the geology of the Lake District. In *Complete Guide to the Lakes by Wordsworth, W. A.* (Hudson and Nicholson, 1842).
13. Underwood, E. A. Lavoisier and the history of respiration. *Proceedings of the Royal Society of Medicine* **37**, 247–264 (1943).

14. Lavoisier, A. L. Sur le charbon. In *Oeuvres de Lavoisier.* **5** (Imprimerie impériale, 1793), pp. 303–310.

15. Rudwick, M. J. S. *Georges Cuvier, Fossil Bones, and Geological Catastrophes: New Translations and Interpretations of the Primary Texts* (University of Chicago Press, 2008).

16. Keynes, R. D. *The Beagle Record: Selections from the Original Pictorial Records and Written Accounts of the Voyage of HMS* Beagle (Cambridge University Press, 1979).

17. Darwin, C. R. *The Structure and Distribution of Coral Reefs.* **15** (D. Appleton, 1842).

18. Darwin, C. R. *Origin of Species* (John Murray, 1872).

7 Physics and Chemistry of Deep Earth

The currently accepted age of Earth is 4.55 billion years, a figure that has not changed significantly for 70 years. From the 1850s to the 1950s, the question of the age of Earth advanced from simply being an airy speculation driven by those geologists willing to allow an indefinitely large amount of past time and those biologists who favored evolution. Unexpected discoveries in physics and chemistry transformed Earth history from intelligent guesswork to a precision science defined by physical laws. This chapter summarizes a century-long pathway of discoveries in physics that would dramatically improve our understanding of the age and properties of the deep Earth, and in particular the deep secrets hidden locked in the carbon atoms in the interior of Earth.

We begin in the Victorian period with William Thomson (1824–1907). He is better known today as Lord Kelvin following his promotion to the British peerage. In 1892, an elderly Queen Victoria ennobled the Scots–Irish mathematical genius on the recommendation of her Prime Minister, the Marquis of Salisbury. The primary motive for this political act was to reward Kelvin's overt opposition to the cause of Irish nationalism. Recognition of his many scientific achievements was a secondary consideration. He was the first scientist to be elevated to the House of Lords, and in Britain he was the most accomplished and influential classical physicist of his time.

In the 1840s, several physical scientists contributed to the concept that energy cannot be created or destroyed in a closed system. Julius Mayer (1814–1878), a German physicist, stated in 1842 that the expenditure of mechanical energy resulted in an equivalent increase

FIGURE 7.1 William Thomson, 1st Baron Kelvin. Bromide print, *c.* 1900 by London Stereoscopic & Photographic Company.
Source: National Portrait Gallery, London. NPG x18984

in the quantity of heat energy when a gas is compressed. James Joule (1818–1889) in Manchester, the center of the Industrial Revolution, came to a similar conclusion when he carried out an experiment using a paddle wheel to vigorously churn water in an insulated can. He, too, found that a given amount of mechanical energy (or work) produced a predictable amount of thermal energy (or heat).

In his mid-career, Kelvin (Figure 7.1) regarded the whole of physics, astronomy and physical geology as legitimately falling within the scope of his inquiries. From his undergraduate days, he had dabbled in the physical problems of geology. His philosophy straddled the science and religion boundary: as a devout Christian, he routinely attended church and chapel. At heart, he remained a creationist, traditionalist and conservative in matters of science and faith. His overarching position can be summarized as an insistence that rational accounts of the natural world must be robustly consistent with the laws of physics. The indefinite timescales that contented James Hutton and Charles Lyell had no place in Kelvin's cosmogony because they ran counter to the laws of thermodynamics. For Kelvin, the engines driving Earth's natural cycles of decay and regeneration could not be perpetual motion machines. Charles Darwin, too, came under his fire for a lack of attention to detail. Kelvin dismissed as absurd Darwin's "one inch per century" for the rate of erosion of the Weald. According to Kelvin, an age of 300 million years could only refer to the time required for the erosion of the mineral calcite that had formed from the minute shells of plankton. It did not encompass the time it took for the chalk to be deposited in the first place. Nor did it address the great majority of geological time, given that the Weald sits at the top of the geologic column. Darwin's simplistic model accounted for only a small fraction of the age of Earth, which must have existed for billions of years. For Kelvin, this amounted to nonsense of a kind that he was anxious to dispel.

Kelvin's fascination with the age of Earth began with his thermodynamic assessment of the age of the Sun, which he assumed to be an enormous hot sphere cooling down as it contracted under its own gravitational attraction in the emptiness of space. That could not last forever. Therefore, the Sun had a finite life, which he had reckoned at 20 million years. Most of the astronomers were content with this astrophysical solution. Kelvin now confronted the uniformitarians and Darwin with hard data: physics was about to trump geology and biology in this debate. The law of conservation of energy was the ace in Kelvin's hand. For the starting position (or the initial boundary condition), he took it that Earth had *solidified* from an originally molten state. His strict mathematical approach demanded simplicity rather than complexity and therefore, to make his analysis tractable, he treated Earth as a rigid *solid* sphere with physical properties that were *uniform* throughout. Cautiously (and wisely as it turned out), he expected his estimates to be valid "unless sources now unknown to us are prepared in the great storehouse of creation."[1] However, one law and three reasonable assumptions, although necessary, were not sufficient to constrain the mathematical model. Kelvin also needed to know the rate at which the internal temperature of Earth increased with depth and the thermal conductivity of the rock in its interior. Neither of these was known when he first turned his attention to the problem, but he made reasonable guesses. In 1864, he proposed that the age of Earth is between 20 million and 400 million years, with 100 million years being the most likely age. As more data became available, Kelvin settled first on 100 million years (1871), but later cut that back to 24 million years in 1895 when new results on the melting temperature of rock became available.[2–4] These estimates did not win any support from the geologists. Although they had no quibbles with Kelvin's mathematics, they disagreed with his modeling assumptions. Matters became heated in the 1890s when biologists became entangled in the tensions between geologists and physicists. Their public debates shifted in emphasis from arguments about the age of Earth to focus instead on the validity of the theory of evolution by natural selection.

The failure of the two camps to agree stimulated John Perry (1850–1920), a former lab assistant and great admirer of Kelvin, to challenge the validity of his underlying assumptions. In 1895, *Nature* published a correspondence initiated by Perry on the age of Earth.[5] Perry questioned Kelvin's assumptions that Earth's interior has a low thermal conductivity and is structurally homogeneous. He politely criticized Kelvin's rigid adherence to the solidity of Earth, pointing out that if the interior were partly fluid it would transport thermal energy by convection much faster than the rate in Kelvin's calculations. Despite Perry's intervention, Kelvin continued to reject the notion of a fluid interior. This was not out of sheer stubbornness. He felt on firm ground because in 1863 his study of Earth's tides had shown that the upper mantle is as rigid as steel.[6] Perry's paper had little impact in terms of citations, possibly because the geologists mostly regarded it as an abstruse tussle between mathematical physicists. Although the idealized calculus of mathematicians had little to attract fieldwork scientists, this did not deter them from standing up to Kelvin's arrogance.

The Reverend Osmond Fisher (1817–1914), who in 1881 published what is certainly the first textbook of geophysics, had pointed out back in 1874 that the great man's calculations concerning the cooling of a solid Earth would be null and void if the interior were in fact fluid, as he believed to be the case. Archibald Geikie (1839–1915), in his 1892 presidential address to the British Association for the Advancement of Science, forthrightly voiced his reservations: "I can hardly doubt ... there must be some flaw in the physical argument." He continued, "Some assumption ... has been made ... which will eventually be seen to vitiate the conclusions ... and allow time enough for any reasonable interpretation of the geological record." In 1897, Kelvin responded robustly to this rising challenge.[7] He resolutely kept to his concept of an Earth cooling through thermal conductivity. Furthermore, he invoked the support of Clarence King (1842–1901), the first director of the US Geological Survey, who in 1893 had settled on a cooling age of 24 million years. All of that provoked a lengthy polemic from Geikie, who sharply criticized Kelvin's aloofness and dismissive attitude when he was confronted with the abundant evidence for Earth's antiquity, compiled over more than a century by geologists and paleontologists. For Geikie, it was the last straw. By 1899, he was so exasperated by the recently elevated peer now lording it over the geologists that he effectively declared independence:

> Are we then to be told that this knowledge, so patiently accumulated from innumerable observations and so laboriously co-ordinated and classified, is to be held of none account in comparison to the conclusions of physical science ... ? These conclusions are founded on assumptions which may or may not correspond with the truth. ... [W]e decline to accept them as a final pronouncement of science on the subject. ... [We] affirm that unless the deductions we draw from that evidence can be disproved, we are entitled to maintain them as entirely borne out by the testimony of the rocks.[8]

The final paragraph of Geikie's address to the British Association for the Advancement of Science called for better international coordination of the observations and data collection by geologists in all countries. He suggested that the International Geological Congress (founded in 1875) "seems well-adapted to undertake and control an enterprise" through a global network of observers undertaking a specific research theme. He listed several topics in Earth history that would benefit from a collaborative approach.

In 1899, Kelvin received another hammerblow, this time from Thomas Chamberlin (1843–1928). Professor of geology at the new University of Chicago, Chamberlin was a prominent and well-respected figure in geological circles and a former head of the glacial division of the US Geological Survey. Chamberlin's early interests in glaciation and the landforms of Wisconsin and neighboring states led him to ask broad questions on the cycles of glacier formation, growth and recession so clearly delineated in these landforms. This, in turn, took him in the direction of the causes of climate change, where his investigation focused on changing levels of carbon dioxide in the atmosphere. Chamberlin issued a powerful rebuttal of Kelvin's methods and assumptions in *Science*. As we shall see in due course, Chamberlin's reasoning was uncannily prophetic. Writing three years after Antoine Henri Becquerel's (1852–1908) discovery of radioactivity, he realized that physics was on the verge of a dramatic revolution. He expressed it like this:

> What the internal constitution of the atoms may be is yet an open question. It is not improbable that they are complex organizations and the seats of enormous energies. Certainly, no careful chemist would affirm either that the atoms are really elementary or that there may not be locked up in them energies of the first order of magnitude ... Nor would he probably feel prepared to affirm or deny that the extraordinary conditions which reside in the center of the sun may not set free a portion of this energy.

Finally, a remark on the popular account by myth-makers claiming that Kelvin had enormously underestimated the age of

Earth because the discovery of those "sources now unknown to us" – *the decay of radioactive elements* – lay in the future. That is not entirely correct. If the known rate of radioactive heating is plugged into Kelvin's calculations, the resulting age of Earth comes out too low. And he was correct in pointing out that Earth is continually losing energy because all geological processes involve irreversible losses of thermal energy. Kelvin's unfortunate error lay with his *physical model* of Earth that did not allow for the *fluidity* of its interior, which meant there could be no transfer of heat by convection. In an accident of history, Perry's powerful argument in favor of a mobile interior continued to be overlooked until the 1960s.[9] In due course, we'll get to the fundamental role that convection in the mantle plays in the deep carbon cycle.

⊙ RADIOACTIVITY RAISES THE STAKES

At the turn of the century, the discovery of radioactivity rapidly found an application in the Earth sciences. The recognition of an abundant source of thermal energy provided a huge stimulus to research on energy transformations in Earth's crust. At the same time, investigation of the physical mechanism of radioactive decay quickly led to the development of powerful new tools for measuring the ages of rocks. These technical developments have proved indispensable for digging into the *history* of Earth.

Successive steps on the road to a great discovery in atomic physics were made by several scientists. This tale of the unexpected begins in 1895, on the evening of the November 8, when, in a totally dark room in Munich, Wilhelm Röntgen (1845–1923) discovered X-rays and their remarkable property of being able to pass through objects that are opaque to ordinary light. On November 22, he photographed the bones, flesh and wedding ring of his wife Anna's left hand (Figure 7.2). Somewhat taken aback by the gruesome image, she exclaimed, "I have seen my own death." This was the first medical X-ray. Within weeks physicians were using X-rays to examine bone fractures and gunshot wounds.

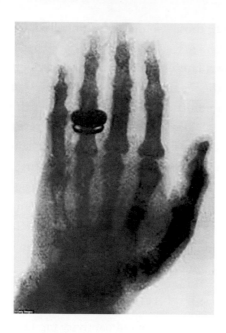

FIGURE 7.2 Historic X-ray of Anna Röntgen's left hand.

Röntgen's chance discovery, which in 1901 earned him the first Nobel Prize in Physics, immediately sparked intense interest among physicists and chemists.[10,11] The production of X-rays by an electrical discharge through a glass vacuum tube seemed to be connected to the strong phosphorescence and fluorescence excited in the glass. Some physicists decided to see whether ordinary matter emitted invisible X-rays, and it was natural that they examined minerals that glowed brightly when exposed to strong sunlight. Henri Becquerel (1852–1908), professor of physics at the Muséum National d'Histoire Naturelle, decided to experiment with uranium salts. He still had some fine crystals of a uranium compound left over from his mineralogical investigations on phosphorescence 15 years back. However, the overcast weather in Paris delayed his intended experiment, so he decided to postpone it until a fine sunny day. He set aside his photographic plates, which were wrapped in black paper to exclude light, together with some crystals of the uranium mineral.

A few days later, he opened a cabinet drawer in the laboratory and took out the plates: "I decided to develop these plates, expecting to find the images very feeble. On the contrary, the plates were blackened with great intensity." Becquerel had expected there would be a faint image owing to a residual amount of phosphorescence that had not quite decayed away. He made further observations to establish the properties of the radiation that had penetrated the black paper and affected the photographic plates. In 1896, his papers announced that the emission was an intrinsic property of the element uranium and entirely independent of phosphorescence. Other physicists mostly did not follow up on Becquerel's discovery, perhaps because they were too busy working on X-rays, which were much stronger than "uranium rays."

⊙ THE CURIES DISCOVER THE SOURCE OF EARTH'S
   INTERNAL HEATING

Becquerel's discoveries intrigued Marie Curie (1867–1934), and they attracted her as a possible field of research for a doctoral thesis. This was at the time when no woman anywhere in the world had been awarded a doctorate for original scientific research. Becquerel became her thesis adviser. Her observations soon confirmed that the emissions of uranium compounds could be an atomic property of the element uranium – a property built into the structure of its atoms. She tested all of the known elements for similar properties and found that only thorium exhibited the same behavior. She invented the term "radioactivity" to describe this feature (Figure 7.3).

What happened next is a deservedly well-known story of heroic proportions. Marie noticed that the mineral pitchblende (or uraninite) was far more radioactive than could be accounted for by its uranium content alone. She conjectured that there must be an additional radio-active element in pitchblende: "The fact is very remarkable and leads to the belief that these minerals may contain an element which is much more active than uranium." She passionately wanted to verify her hypothesis as quickly as possible in order to establish the priority for such a discovery. Pierre Curie (1859–1906) pitched in, aiding his wife in the search for the unknown substance. Pitchblende is a

FIGURE 7.3 Pierre and Marie Curie, 1895, in the garden of the house of Pierre's parents. The young couple were on their bicycling honeymoon. Photographer Albert Harlingue.
Source: Wikimedia Commons

complex mineral made up of combinations of about 30 chemical elements. Only a tiny amount of the unknown radioactive substance would be present, so they were seeking the proverbial *"aiguille dans une botte de foin"* – needle in a haystack. They set to work by using standard, but demanding, chemical procedures to separate the different substances. The couple discovered that two of their fractions, one dominated by bismuth and a second containing mostly barium, were strongly radioactive. This could be explained only by the presence of two new elements, both of which must be highly radioactive. In July 1898, Marie suggested the name "polonium" for the new element that shared many of the chemical properties of bismuth. A second paper, in December 1898, explained that the barium fraction contained another new element, radium. These discoveries were a wake-up call for the apathetic physics community that had preferred the research frontier Röntgen's X-rays had opened up rather than Becquerel's uranium rays, which they suspected of being an electromagnetic phenomenon.

From 1899, the number of published papers on radioactivity increased markedly.[12] In order to convince the scientific community of the existence of polonium and radium and to establish the physical and chemical properties of both, Marie set out to isolate them from bismuth and barium, but the difficulties she faced were truly enormous. The Austrian government donated several tons of pitchblende, which was a waste by-product of the mines at Joachimstal that Agricola had known so well. It took three years of sweated labor to isolate just one-tenth of a gram of pure radium chloride from thousands of kilograms of ore. But that was sufficient for Marie to measure radium's atomic weight of 225. She presented this work in her doctoral thesis on June 25, 1903: her examination committee (three men) lauded her findings as the greatest scientific contribution ever made in a doctoral thesis – which, arguably, is still true today.

Three months earlier, on March 16, 1903, Pierre and his young lab assistant Albert Laborde had broken the news in *Comptes Rendus* of a dramatic discovery that launched a radically different era for the development of theories about the internal workings of Earth. Their paper, whose title translates in English as "On the heat released spontaneously by radium salts," made the startling claim that in less than one hour, radium produces sufficient heat to melt its own mass of ice. Hour after hour. Day after day. Year after year. For centuries. Radium is a million times more radioactive than uranium. Pierre realized straightaway that the stupendous energy being released by radium would provide an important source of heating in Earth's interior. In December 1903, the Curies and Becquerel were awarded the Nobel Prize in Physics. The citation for the award avoided specific mention of their discovery of polonium and radium: the chemists on the nominating committee insisted that the Curies might in the future deserve the Nobel Prize in Chemistry for discovering those elements, which Marie alone received in 1911. Her husband had been tragically killed instantly on April 19, 1906, when he was struck down in the street by a horse-drawn wagon weighing some six tons. In 1995, the remains of the Curies were reinterred at the Panthéon.

On August 16, 2017, Professor Hélène Langevin-Joliot, granddaughter of Pierre and Marie, gave a Plenary Lecture to the 27th Goldschmidt international conference on geochemistry that convened in Paris. She titled her presentation "Pierre and Marie Curie and Radioactivity," and in her conclusion she emphasized: "Radioactivity, a name coined by Marie Curie, became a major research field in the decades that followed [her discovery]. The huge amount of thermal energy produced by radioactive elements present inside the Earth has played an important role in the Earth's internal dynamics over its billions of years of history."

⊙ RADIOACTIVE CLOCKS LOCKED IN ROCKS

The discovery of radioactivity provided a novel and robust way to measure the age of ancient rocks that made all other geological dating methods obsolete. The leaders in this new quest were two early-career scientists: Ernest Rutherford (1871–1937) and Frederick Soddy (1877–1956). The foundations for their collaboration began in England, in Cambridge and Oxford, respectively.

Cambridge first admitted research students from other universities in 1895. The first two such students taken on by the Cavendish Laboratory were Rutherford, a physics graduate with three degrees from the University of New Zealand, and John Townsend (1868–1957), a mathematician from Trinity College, Dublin. Both were supervised by Joseph John Thomson (1856–1940), who had discovered the electron early in 1897. Rutherford had already developed a keen sense of what was new and exciting in physics. For his MA dissertation he initially investigated the effects of X-rays on gases but, on learning of Becquerel's discoveries, he switched his attention

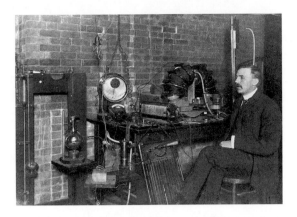

FIGURE 7.4 Ernest Rutherford in his laboratory at McGill University, Montreal, Canada, in 1905.
Source: Library of Congress, Prints & Photographs Division, LC-B2-1234

to the radioactivity of uranium. In 1898, he discovered that radio-active atoms emit two types of rays that differ from X-rays in their penetrating power. Later that year, Thomson recommended Rutherford for appointment to the Macdonald research professorship of experimental physics at McGill University, Montreal, Canada (Figure 7.4).

In May 1900, Soddy, of Merton College, Oxford, arrived in Montreal in a junior faculty position at the chemistry department. He and Rutherford first met in September of that year, and over the next 18 months the pair engaged in a frenzy of experiments on radio-activity that led to 9 publications.[13] They were the first to label the three distinct types of radiation as alpha, beta and gamma. Alpha radiation consists of the nuclei of helium atoms; beta rays are electrons; and gamma rays are electromagnetic radiation similar to X-rays but even more energetic. From close examination of thorium, radium and uranium they concluded that radioactive elements are in a state of transformation, as a result of which a number of new substances are produced, entirely distinct from the parent elements in their chemical and physical properties. They realized that through the emission of alpha or beta radiation radioactive elements can spontaneously change into other elements. This spontaneous decay process continues in a chain of emissions until a stable atom is formed. They proposed that there were two radioactive decay series: one beginning with uranium and ending in lead; and the second starting with thorium and finishing with lead. It was, Rutherford and Soddy boasted, the transmutation of elements that had eluded alchemists for thousands of years.

This careful work by Rutherford and Soddy proved that the enormous amount of thermal energy emitted is a consequence of the release of alpha radiation. By 1903, Rutherford was convinced that the atoms of radioactive elements, and possibly all elements, encapsulated stupendous amounts of latent energy, and that the source of that energy was a process of subatomic change. In plain English, Rutherford saw that Kelvin's law of conservation of energy implied that the energy of radioactive decay must be an intrinsic property of radioactive atoms. Chamberlin's declaration in 1899 that "the internal constitution of the atoms is as yet an open question" had turned out to be a sensible precaution.

Rutherford now turned to popularizing the new research field of radioactivity. Early in 1904, he delivered a Friday Evening Discourse at the Royal Institution in London, where he explained to an attentive general audience that radium in Earth's crust liberated stupendous amounts of energy. The implications for biology and geology were immense because of the great increase in the time available for evolution by natural selection. Casting his eye over the half-dark room, Rutherford (then aged 32) spotted the recumbent figure of Lord Kelvin (who was pushing 80). Just when Rutherford asserted that Earth could no longer be treated as a cooling sphere because of the huge amount of energy being released by decaying radium, the peer perked up. But it was just a momentary lapse. He continued to reject the claims that radium could perpetually emit heat, arguing instead that some external agency, unknown at the time, must be involved. By this time, the stubborn insistence of an old man looked rather ridiculous: most physicists accepted that radioactivity was in some way responsible for Earth's heat, and, indeed, that of the Sun. Kelvin never published a retraction. Meanwhile, most field geologists remained detached from the tumult inside the physics laboratories, where a new atomic physics of photons, rampant radioactivity, special relativity, $e = mc^2$ and invisible quantum curiosities seemed entirely disconnected from the solid Earth.

Rutherford and Soddy noted that the degree of radioactivity in a sample is proportional to the number of radioactive atoms present in the sample. This means that the activity of a particular element falls by a half in a characteristic period of time called the half-life, which varies from one element to another. For example, the half-life of the most stable type of radium is 1600 years, whereas the most abundant type of uranium has a half-life of almost 4.7 billion years. In mathematical parlance, we can say that radioactive decay declines exponentially: the amount of radioactivity in a sample decreases at a rate proportional to its current value. Rutherford regarded the very long half-life of uranium as a natural clock built into radioactive substances. He already knew that a by-product of the disintegration of uranium was the production of helium, which the chemist Sir William Ramsay (1852–1916) had uncovered trapped in uraninite in 1895. Therefore, the proportion of helium left trapped in a rock could be a proxy for the time that had elapsed since the formation of the rock.

In his Silliman Memorial Lecture given at Yale University in 1905, Rutherford made a clear statement about the potential of using radioactivity to measure the age of rocks. In his own words:

> ... it should thus be possible to determine the interval required for the production of the amount of helium observed in radioactive minerals, or, in other words, to determine the age of the mineral.

He showed two mineral samples containing uranium, saying they were probably 500 million years old based on the amount of helium they contained, and he floated the idea that perhaps calculations based the percentage of lead in radioactive samples rather than helium would give a better result. He accepted that some helium had out-gassed, so he felt that his ages were underestimates.

All of this intrigued Bertram Boltwood (1870–1957), a young graduate of chemistry in the audience, who had just received an offer of an assistant professorship at Yale. He decided to make his own investigation of the disintegration of uranium. The following year, by looking carefully at the ratios of the proportions of uranium and lead in rocks, he discovered that there was less uranium and a higher proportion of lead in the older rocks. He deduced that lead is the final stable element to form when uranium decays into a succession of radioactive elements. Unlike helium, lead cannot escape easily from rocks, so measuring the amount of lead in a sample should provide a more reliable age than measuring helium. With Rutherford's encouragement, Boltwood began to apply his new method to 43 rock samples collected from 10 different locations. In 1907, he published his findings: the rocks had formed between 410 and 2200 million years ago. Despite being the pioneer of the uranium–lead method of geochronology, which was to become his outstanding scientific achievement, he remained firmly committed to analytical radiochemistry, and made no further contributions to geoscience. At this stage, the torch passed from New Haven to London.[14]

⊙ ARTHUR HOLMES DATES EARTH

In London, at the same time that Boltwood was working on geochronology, Robert Strutt FRS (1875–1947; and from 1919 the fourth Lord Rayleigh) also followed up Rutherford's hint that lead would probably be a more trustworthy proxy for age than helium. This set the scene for Arthur Holmes (1890–1965) (Figure 7.5) to illuminate a pathway to the deep history of Earth by his pioneering use of the radiometric dating of minerals. A humorous and open-minded person, he entered Imperial College London in 1907. On graduating in geology and physics in 1910, Holmes commenced postgraduate research on the application of radioactivity to problems of the geological timescale. Holmes had first become fascinated by this topic as a youngster when

FIGURE 7.5 Arthur Holmes.
Source: University of Edinburgh
School of GeoSciences

he noticed on the first page of his parents' bible the marginal note giving the date of the Creation as 4004 BC. "I was puzzled by the odd '4'. Why not a nice round 4000 years? And why such a recent date? And how could anyone know?"[15]

At his laboratory bench, Holmes had the good fortune to be guided by Strutt, who had recently been appointed to the professorship of physics at Imperial.[16] He embarked on what became his lifetime's quest "to graduate the geological column with an ever-increasingly accurate time scale."[15] For his very first uranium–lead analysis he carefully selected a Devonian igneous rock (nepheline syenite) from Norway that contained 17 different radioactive minerals. He worked with 100 kg to recover 0.3–2.0 g of each. This allowed him to make internal checks on the uranium–lead ratios, after which he discarded nine as being unsuitable because they evidently had already contained lead when they formed. The remaining eight yielded an age of 370 Ma (370 million years). The geological age of the rock was Devonian, and so Holmes' 370 Ma became the first radiometrically defined reference point for that period. For Boltwood's samples, Holmes obtained radiogenic ages for eight rock minerals ranging from 340 Ma (Carboniferous) to 1640 Ma (Precambrian), and the resulting tabulation of absolute radiometric age showed a remarkable correlation with relative geological age. He put it like this in his 1911 paper: "Indeed, it may confidently be hoped that this very method may in turn be applied to help the geologist in his most difficult task, that of unraveling the mystery of the oldest rocks of the earth's crust."[17] Holmes was the first Earth scientist to combine radiogenic ages with geological formations in order to create a geological timescale. In April 1911, Strutt presented this paper on Holmes' behalf at a meeting of the Royal Society. Although the geologists showed some interest, as ever many were wary of newfangled procedures peddled by physicists, questioning their assumptions about the decay rate of uranium.

Holmes was not present at the reading of his paper because he was overseas; having found it impossible to live in London on his fellowship stipend of £60 a year, he was desperately in need of hard cash. That's why he signed on as a professional geologist for a risky expedition, on the payroll of a mining company, to carry out geological surveying and prospecting in Mozambique (then a province of Portuguese East Africa, with minimal European settlement, mostly by missionaries). Although the rock-hounds unearthed nothing of economic benefit while sweating in the tropical jungle, the trip was fruitful for young Holmes, who collected, for later analysis back home, samples of Precambrian rocks and zircons, all of good quality for age determinations. There was a setback, however. At the end of his six-month contract, while homeward bound, he was struck down by malaria, and then became gravely ill from an often-fatal complication: blackwater fever. The nuns caring for him at a missionary hospital prematurely telegraphed notice of his death to London. Fortunately, he made an extraordinary recovery, and in November 1911, he arrived back at Southampton in style, on board the express passenger liner RMS *Edinburgh Castle*. He suffered bouts of malaria for years afterwards. In 1912, Imperial gave him an early-career

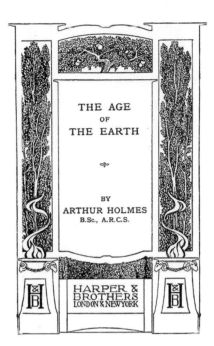

FIGURE 7.6 Title page, *The Age of the Earth*, 1913.

position as a demonstrator, which allowed him to work on the petrographic samples from East Africa. When World War I broke out in August 1914, he immediately volunteered, but fortunately for his sake – and for the future of geosciences – he was among the 40 percent of young men rejected for medical reasons. When he took up the demonstrator post at the age of 22, he put pen to paper to promote the radiogenic dating of rocks. His famous book, *The Age of the Earth* (Figure 7.6), published in London and New York by Harper and Brothers in March 1913, opens with these now famous lines:

> It is perhaps a little indelicate to ask of our Mother Earth her age, but Science acknowledges no shame and from time to time has boldly attempted to wrest from her a secret which is proverbially well guarded.

He estimated the oldest age of Archean rocks to be 1600 Ma. Clearly, Earth was older than this, but he stopped short of speculating by how much. He preferred to wait for more data, which were not long in coming.

⊙ EARTH'S DYNAMIC DEEP INTERIOR

Holmes was only 23 when his popular science book was published, and it took well over a decade before he was recognized as a world authority on the subject. In fact, he continued to face formidable opposition from established geologists who remained entrenched in the belief that Earth was 50–150 Ma. In the final chapter in *The Age of the Earth*, Holmes expressed his regret that his pioneering approach "had been shaken by advocates of the geological method of attack." The geologists had not welcomed the intrusion of the marvelous properties of radium. Curiously, having spent a century arguing that millions were needed in the banks of time, they were now backing off when faced with billions. Holmes felt "we are at the parting of the ways" because "there exists a serious discrepancy," pointing to "a flaw in the underlying assumptions of one or the other or both of the methods." The pendulum had swung too far. The skepticism drifted on for about 15 years. In 1924, Frank Wigglesworth Clarke (1847–1931) of the US Geological Survey commented in a geochemical data review that astronomical data such as the age of the Sun, chemical denudation rates and paleontological evidence all pointed to ages of Earth in the range 50–150 Ma: "The high value found by radioactive measurements are therefore to be suspected until the discrepancies shall have been explained."[18]

In searching for that "flaw in the underlying assumptions," Holmes felt that the only item he could reasonably question was the assumption of uniformity that was present in both the geological and the radioactive methods. In the former case, this took the form of the well-worn doctrine that all change must be gradual, and in the latter case that the rate of disintegration was the same in the past as at present. Naturally, with the enthusiasm of youth, he pointed the finger of suspicion at the geological ages. Could it be that they were too short because denudation (erosion) had taken place more slowly in the past, and therefore the rocks had endured for longer?

In a daringly speculative move, Holmes ditched the dogmatic approach to uniformity, advocating instead a modern interpretation based on a more liberal conception of evolution. The Curies had shown that internal heating occurred in Earth's interior. This would allow the development of Earth as a cyclical phenomenon. If Earth was not cooling down, then the deposition of sediments, marine transgression and recession (the rise and fall of sea level relative to land) and igneous activity (volcanic activity and earthquakes) could all be seen as rhythmic phenomena reflecting a common factor: "Earth movement." For good measure, he even threw in the weathering effect of rain, which is enhanced in proportion to the dissolved carbon dioxide and which "in turn is conditioned by the prevalence of volcanism."

To the mosaic of methods, ideas, hypotheses and speculation that scientists had used in their search for the truth about Earth's antiquity, Holmes brought *skepticism* in place of disbelief – a timely reminder that scientific fact and interpretation are never absolutely true. In antiquity, the Skeptics held that there are no adequate grounds for certainty as to the truth of any proposition. In science, skepticism may bring great rewards and recognition in the form of medals and prizes: as we have seen, it did so for the Curies, Rutherford and Holmes. We shall return to the remarkable career of Arthur Holmes in the next chapter. But before that, a brief coda concludes the story of his lifelong interest in dating Earth and the contributions to deep Earth science made by the next generation of geoscientists.

⊙ ISOTOPES EXTEND THE GEOLOGICAL TIMESCALE

First, we shall catch up on developments in radiochemistry. The papers that Rutherford and Soddy published on their work at McGill inspired other chemists to search for more radioelements in the decay chains that terminated in lead. By the time Holmes was correcting the proofs of *The Age of the Earth*, the motley collection of radioelements numbered almost 40. On page 190, Holmes helpfully provided a block diagram of the uranium, actinium and thorium families, together with the half-lives of the members. Many of the decay products had such short half-lives (hours, minutes, seconds even) that they could not be distinguished or characterized by mass, and their alpha and beta emissions showed no apparent patterns. What a mess. It was impossible to slot all of these elements into the periodic table. However, wet chemistry led to an elegant solution of this intellectual puzzle when chemists noticed striking chemical similarities between a few of the radioelements.

The full story is an unnecessary complication in this account of deep carbon science, but it is worth noting that the discovery did lead to profoundly important developments in geoscience. Soddy first separated elements with identical chemical properties but different half-lives into groups, and then he proposed that the species that were chemically identical should share the same location in the periodic table, despite having different atomic masses. He outlined this concept during a social dinner party at the home of his wife's parents. One of those dining, Margaret Todd (1859–1918), a pioneering medical doctor and novelist, suggested that atoms sharing the same place in the periodic table should be named *isotopes*, from the Greek words *iso* for "same" and *topos* for "place." He did exactly that in his 1913 *Nature* paper. The citation for his Nobel Prize in Chemistry (1921) was for "his contributions to our knowledge of the chemistry of radioactive substances, and his investigations into the origin and nature of isotopes." The discovery of isotopes made the task of calculating uranium–lead ratios more complex, but with the promise of greater precision. In the second edition of *The Age of the Earth* (1927), Holmes had upped his estimate for Earth's age to 1600–3000 Ma, although with considerable uncertainty.

Meanwhile, Francis Aston (1877–1945) of Cambridge had constructed a mass spectrograph with sufficient resolution to identify 212 stable isotopes, a superlative achievement for which he was awarded the 1922 Nobel Prize in Chemistry. He discovered the lead isotopes 206, 207 and 208, which confirmed Soddy's hunch that the three uranium radioactive series terminated at three different isotopes of lead. In 1927, by using uranium–lead ratios, Holmes and Robert Lawson significantly improved the detail of the geological timescale out to 1336 Ma. From Aston's data, in 1929, Clarence Norman Fenner (1870–1949) and Charles S. Piggot (1892–1973) of the Carnegie Institution of Washington, DC, made the first calculations of a mineral's age using mass spectroscopy of lead isotopes. Isotope dating was born, and it would transform geoscience. The predominant uranium isotope is ^{238}U with a half-life of 4.47 billion years. In 1935, Arthur Dempster (1886–1950) at the University of Chicago discovered its minor partner ^{235}U, which has a half-life of 700 million years and a natural abundance of just 0.7 percent of that of ^{238}U. The importance of these advances for geochronology was that the ^{238}U decay chain terminates with ^{206}Pb, while ^{235}U ends with ^{207}Pb. Primordial lead, which comprises the stable isotopes 204, 206, 207 and 208, is the result of nucleosynthesis that took place variously in dying stars, mergers of neutron stars or supernova explosions. In Earth's crust, all lead-204 is primordial, whereas the abundances (or proportions) of lead-206, lead-207 and lead-208 are increasing because they are the decay products of the three radioactive series. Therefore, two decay routes provided chronometers running on different timescales (Figure 7.7).

Through the 1920s and 1930s, radiometric measurements of rocks steadily pushed Earth's age back. Even though radiometric dating methods were still imperfect and subject to a good deal of uncertainty, scientists were routinely measuring Earth's age in billions of years rather than millions. In the USSR in 1942, E. K. Gerling (1904–1985) of the Radium Institute compared the isotopic proportions of radiogenic lead-206 and lead-207 with that of the primordial lead-204. Gerling obtained an age of 3950 Ma for a lead ore sample from Ivigtut, Greenland. However, his paper was published in Russian in the middle of World War II, so his success went unnoticed until it was translated several years later. In the postwar recovery period, Fritz

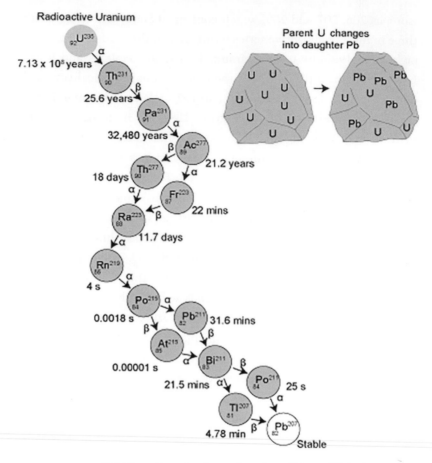

FIGURE 7.7 The radioactive decay chain from ^{235}U to ^{207}Pb, the application of which to geochronology significantly improved the accuracy of the longest dates.

Houtermans (1903–1966) in Göttingen and Holmes in Edinburgh vigorously pursued essentially the same path as Gerling had done, and each published their results in 1946 and 1947, respectively. The final sentence of Holmes' paper published in *Nature* on January 27, 1947, is:

> It is, therefore, concluded that, on the evidence at present available, the most probable age of the Earth is about 3350 million years.[19]

These age estimates for Earth led to a historical problem for astronomers, who were also in the dating game – in their case, to measure the age of the universe. Edwin Hubble (1889–1953), a staff member of the Carnegie Institution, was the greatest observational astronomer of his generation, with privileged access to the 100-inch telescope at Mount Wilson Observatory, California. In the mid-1920s, he became the first astronomer to measure the *distances* of the galaxies. His data, when combined with the *velocities* of the galaxies that others had already measured, enabled theoretical astronomers to calculate an age for the universe, as measured from the onset of the Big Bang: 1.8 billion years, rather less than the 2 billion that Holmes claimed in the mid-1920s.

Astronomers were not too bothered by this discrepancy: for them, the geological age was unacceptably indeterminate. The standing of the two protagonists in their respective communities – Hubble (aged 40) and the stargazers on the one side, with Holmes (aged 39) and the rock-hounds on the other – did not help the dispute. Hubble surveyed the universe with the world's largest telescope, and he already commanded huge respect as the person who had discovered the bewildering vastness of the cosmos in 1922–1923. And now theorists were citing his data obtained with an amazing telescope in terms of an expanding universe with an age of 1.8 billion years. By contrast, Holmes had nothing like Hubble's reputation, and the astronomers were unfamiliar with uranium–lead dating.

The stark contrast between the astronomical and the geological timescales became a driver of new thinking in cosmology. Even Albert Einstein began to fear for the future of cosmology:

> The age of the universe … must certainly exceed that of the firm crust of the Earth as found from radioactive materials. Since determination of age by these minerals is reliable in every respect, cosmological theory … would be disproved if it were found to contradict any such results. In [which] case I see no reasonable solution.[20]

By 1948, the age controversy was critical, largely because 3350 Ma as determined by lead isotopic abundances had become widely accepted as a *lower* limit for the age of Earth. Three cosmologists went so far as to reject the law of conservation of mass: Fred Hoyle, Hermann Bondi and Thomas Gold postulated a universe that had always existed in the past and will always exist in the future. They waved a magic wand to make matter appear from nowhere to fill the void left by expansion. In this picture, the rate of expansion of the universe has nothing to say to us about the age of the universe, which is infinite. The theory was refuted in 1965 when radio astronomers detected the fossil radiation from the Big Bang. The conflict of ages was resolved in 1952 when Walter Baade (1893-1960) of the Mount Wilson Observatory announced a new calibration of the cosmic distance scale. He had spent several years making observations of the variable stars that astronomers used as "standard candles" for calibration. Newspapers of the time seized the chance to use catchy headlines such as "universe doubled in size." The age of the universe jumped to 4 billion years, which aligned nicely with the radioactive ages. Problem solved.

Clair Patterson (1922–1995) now graces our story. His research centered on the geochemistry of lead in rocks, meteorites, water and the atmosphere. Houtermans suggested that the initial mix of lead isotopes could be found by examining iron meteorites. As we saw in Chapter 2, meteorites come from asteroids that have remained largely unchanged since the early days of the solar system, so they should preserve a record of the solar system's initial composition. Patterson took up Houtermans' idea and measured the lead isotopes in an iron sulfide mineral sample (FeS) taken from the Canyon Diablo meteorite, which had crash-landed in Arizona some 60,000 years ago. The amount of uranium in this sample was so small that its lead isotope ratios must have stayed almost constant since the meteorite's parent asteroid formed. By assuming that this mixture really was primordial, Patterson could now calculate the age of the parent asteroid: it was 4.55 billion years old. Soon afterward, Houtermans made a similar measurement and arrived at an almost identical conclusion. This left one vital question: Did Earth and the asteroid have identical lead ratios to begin with? To test this, Patterson examined the lead found in deep-sea sediments from the floor of the Pacific Ocean, arguing that the mixture of lead isotopes in these sediments should be similar to Earth as a whole. The sea sediment results agreed precisely with the data from five separate meteorites. This meant that Earth and the parent asteroids of these meteorites had all formed from a common reservoir of material with the same mixture of lead isotopes. This yielded an age of 4.55 billion years for Earth, a 10-percent increase on Gerling's 3.95 billion years for the oldest rock. All subsequent research on the isotopic and radiometric ages of rocks, meteorites and lunar samples have confirmed the results that Patterson and Houtermans arrived at in 1953: 4.55 billion years. Henceforth, the age of Earth and the time available for geological and biological global cycles were no longer sources of conflict in the worldwide community of geoscientists.

REFERENCES

1. Thomson, W. On the age of the Sun's heat. *Macmillian's Magazine* **5**, 288–293 (1862).
2. Thomson, W. On geological time. *Transactions of the Geological Society of Glasgow* **3**, 321–329 (1871).
3. Thomson, W. The age of the Earth. *Nature* **51**, 438–440 (1895).
4. King, C. The age of the Earth. *American Journal of Science* **45**, 1–20 (1893).
5. Perry, J. On the age of the Earth. *Nature* **51**, 582–585 (1895).
6. Kelvin, W. T. On the rigidity of the Earth; shiftings of the Earth's instantaneous axis of rotation; and irregularities of the Earth as a timekeeper. *Philosophical Transactions of the Royal Society of London* **153**, 573–582 (1863).
7. Kelvin, W. T. The age of the Earth as an abode fitted for life science. *Annual Address (1897) to the Victoria Institute* **9**, 665–674 (1897).
8. Murray, J. Geikie, Archibald. Presidential Address given in Dover to the Geology Section of the British Association for the Advancement of Science, Saturday, September 16, 1899. *Report of the British Association for the Advancement of Science* (1900).

9. England, P., Molnar, P. and Righter, F. John Perry's neglected critique of Kelvin's age for the Earth: a missed opportunity in geodynamics. *GSA Today* **17**, 4–9 (2007).

10. Rutherford, E. *Radioactive Substances and Their Radiations* (Cambridge University Press, 1913).

11. Holmes, A. *The Age of the Earth* (Harper and Brothers, 1913).

12. Malley, M. The discovery of atomic transmutation: scientific styles and philosophies in France and Britain. *Isis* **70**, 213–223 (1979).

13. Fleck, A. Frederick Soddy, 1877–1956. *Biographical Memoirs of Fellows of the Royal Society* **3**, 203–216 (1957).

14. Kovarik, A. F. Biographical memoir of Bertram Borden Boltwood 1870–1927. *National Academy of Sciences* **14**, 69–96 (1929).

15. Lewis, C. *The Dating Game: One Man's Search for the Age of the Earth* (Cambridge University Press, 2002).

16. Dunham, K. C. *Arthur Holmes, 1890–1965.* **12** (Biographical Memoirs of Fellows of the Royal Society, 1966).

17. Holmes, A. The association of lead with uranium in rock-minerals, and its application to the measurement of geological time. *Proceedings of the Royal Society of London* **85**, 248–256 (1911).

18. Dalrymple, B. G. The age of the Earth in the twentieth century: a problem (mostly) solved. *Geological Society, London, Special Publications* **190**, 205–221 (2001).

19. Holmes, A. A revised estimate of the age of the Earth. *Nature* **159**, 127–128 (1947).

20. Einstein, A. *The Meaning of Relativity* (Princeton University Press, 1945).

8 Confronting the Continental Drift Conundrum

⊙ "WHAT WE ARE WITNESSING IS THE COLLAPSE OF THE WORLD"

At the beginning of the twentieth century, conventional geological authority reposed peacefully on the bedrock of uniformitarianism. The received wisdom held that geological processes had changed little across time and that there was no evidence of sudden large-scale changes. Furthermore, the geology of the ocean floor was not considered to be an area of fruitful inquiry. The geological continuity of strata and mountains across deep oceans was explained by the cooling and contracting Earth. Most of the European geologists had accepted the framework of the evolution of Earth's rocky crust that Europe's leading geologist, Eduard Suess (1831–1914), had developed from the mid-1880s.[1] According to his model, parts of paleocontinents had sundered and sunk to form ocean bottoms. He argued that horizontal movements of the lithosphere, rather than vertical uplift, created mountain ranges by thrust faulting and folding. He claimed in 1885 that as a result of the contraction of Earth, "What we are witnessing is the collapse of the world."[2] His was a fixist theory, because continents did not move laterally. Lord Kelvin, the champion fixist and grand overlord of physics, continued to insist on the rigid application of physical theories, such as thermodynamics and the mechanics of solids, as well as plain common sense, whenever he was confronted by anyone who angled for a more flexible approach to the science of the solid Earth.

The American geologist and science writer James Lawrence Powell has identified three highly significant concepts of geology during the twentieth century.[3] These are deep time, the building of our planet and the solar system by myriads of chance collisions of objects in space and the interactions of moving plates and continents that account for almost all of Earth's surface features. These three concepts each constituted revolutions in thinking, even if it did not first appear that way. Each paradigm shift has been profoundly important for developing our understanding of the geochemistry and geophysics of the deep Earth, none more so than the momentous evolution of tectonic theory since the early twentieth century. In this chapter, we will meet a number of very remarkable scientists whose discoveries were to have major implications for understanding the dynamics of Earth's interior.

Our historical prelude begins with Gerardus Mercator (1512–1594), the Flemish geographer and mapmaker who introduced the eponymous cylindrical map projection in 1569. It rapidly became the standard for nautical purposes, because navigators could use it to plot a straight-line course to follow by magnetic compass (i.e. rhumb line) rather than fiddling around with great circles where the bearing kept changing. Mercator made his fortune by selling large wall maps, terrestrial and celestial globes and his geographical atlas of maps published in 1595. Famous as Mercator is, another scholar from the golden age of the Dutch school of mapmaking, Abraham Ortelius (1527–1598), a geographer and scholar of classical antiquity, had a greater impact than Mercator on scientific thought. Ortelius published the first modern atlas (*Theatrum Orbis Terrarum*) in 1570, and in a passage added to his third edition he made two bold observations. First, he noted the way in which the coasts of Europe and Africa seemed to mesh "with the recesses of America." Second, he speculated that "the island Atlantis or America" was "torn away from Europe and Africa, by earthquakes and flood."[4]

> But the vestiges of the rupture reveal themselves, if someone brings forward a map of the world and considers carefully the coasts of the three aforementioned parts of the Earth, where they face each other – I mean the projecting parts of Europe and Africa, of course, along with the recesses of America.

The breakthrough made by Ortelius anticipated the work of the philosopher and statesman Francis Bacon (1561–1626). In his *Novum Organum* ("New Instrument," 1620), the book on which his reputation as a great scientist rests today, Bacon set out a new methodology and philosophy for doing scientific investigations. Bacon briefly mentions the pleasing symmetry of the Old and New World coasts. However, he failed to perceive the congruity of the continents, a remarkable omission that has led modern scholars to question the myth-makers' claims that Bacon was a founder of continental drift.[5] A French prior, François Placet, suggested in a memoir in 1668 that the Old and New Worlds had separated during the biblical Flood. In 1749, the paleontologist Buffon also missed the congruence of coastlines when remarking on the similarities of their fossil life. The continental drift speculations of the polymath Benjamin Franklin and geographer Antonio Snider-Pellegrini (1802–1885) have likewise been downgraded to harmless speculation rather than elevated to brilliant foresight. By contrast, after being overlooked for almost four centuries, Ortelius' star is in the ascendancy, and he now assumes his rightful place in the history of continental drift theory. Why did his elevation take so long? Perhaps it was because Ortelius incorporated his ideas as a single entry in a vast work of reference, where it languished unnoticed until the mid-eighteenth century.

⊙ ALFRED WEGENER SHIFTS THE WORLDVIEW

Alfred Wegener (1880–1930) led a truly adventurous and tragic life in science, during which he upended geology, although no one ever thanked him for doing so. During his short life, hardly anyone took his hypothesis seriously. By the 1950s, his name was but a footnote in elementary textbooks of geology. Falsely, he was written about as a footloose German meteorologist who had blundered by accident into the theory of continental drift while leafing through a new atlas that had maps of the continental margins. After publishing his first outline of the hypothesis of continental displacement, he battled with skeptics for 15 years before succumbing to a terrifying death while exploring the forbidding Greenland ice cap. Wegener is a classic example of the novelist's "hero," a protagonist who fearlessly embarks on daring deeds, in his case the pursuit of better scientific data. In 1905, while making meteorological investigations, he and his older brother Kurt Wegener (1878–1964) broke the world record for a continuous balloon flight by remaining airborne for more than 50 hours. During the first of his four expeditions to Greenland, three of his colleagues died in a wilderness of ice during a challenging excursion by dogsled. On his second foray in 1908, he and a colleague became the first men to overwinter on Greenland's inland ice – in the spring thaw, a Lutheran clergyman visiting a remote congregation rescued them from a fjord in the nick of time, after they had eaten their last pony and dog! The scientific drive behind Wegener's missions to Greenland was the prospect of future transatlantic flights between Europe and North America, which would require an improved knowledge of Greenland's weather. He was one of the last true natural philosophers, by which I mean a person educated in the late nineteenth century whose interests ranged over climatology, geology, geophysics, glaciology, lunar and planetary astronomy and meteoritics, as well as atmospheric and cloud physics.

It was in September 1902 that Wegener began his graduate studies in Berlin, where he signed up for Max Planck's (1858–1947) lectures on thermodynamics, an experience that profoundly shaped his academic career. Planck's philosophical insistence on caution and modesty in physical theory, together with an indifference to causes, characterized all of Wegener's life in science.[6] When his spare-time musings turned to continent drift, he opened up to his closest scientific colleague, the Russo-German climatologist Wladimir Köppen (1846–1940), the founder of modern climatology and meteorology. Alfred would soon join Wladimir's family by marriage to Else Köppen (1892–1992) on his return from Greenland. Wegener wrote to Köppen in December 1911 that his hypothesis of continental displacement was not a fantasy, but rather a recasting of existing data into a new form that replaced complexity and contradiction with simplicity and harmony. As he developed this hypothesis, from 1912 onward, the data he cited included the similarity of the biota on the different continents, the similarities in their fossil records, the continuity of mountain chains and strata (including the carboniferous Coal Measures) across the oceans and paleoclimatological evidence of former glaciations in Australia, South Africa and India. At this time, the hypothesis that extensive land bridges had formerly linked continents, thus enabling global diffusion of plants, animals and humans, continued to have strong support. This was just one of the existential assumptions facing Wegener's challenge. He had explained the continuities by the continents splitting and then drifting apart to form the ocean basins, but there was no solid evidence for fluid continental motion.

Wegener's startling vision first became public on January 6, 1912, when he contributed to a session of the (German) Geological Association meeting held at the Senckenberg Natural History Museum in Frankfurt. The audience was unimpressed by the young meteorologist. Although he displayed admirable skills of exposition, these did not entitle him to set the continents adrift, particularly as he had only been studying the geological literature for about four months![7] He sent the text of his talk for publication by the Geological Association: its unambiguous title was: "The Origin of Continents."[2] The paper's rhetorical structure follows the deductive style of argument he had learned from Planck, but this held no appeal for practical geologists. In it we find his first use of the expression *"die Verschiebung der Kontinente"*– continental displacement (i.e. drift).

Wegener argued that all continents were once conjoined as Pangaea, which was a single landmass stretching from pole to pole that began to break up about 200 million years ago. Until the mid-1970s, a misconception persisted in textbooks that Wegener had used the modern-day continental shorelines for his fit. In fact, he used the 200-m isobaths as the clear boundary between the continental shelf and the abyss. A longer version of the paper, cogent and replete with strong geological evidence, came out in three instalments in 1912.[8]

Once again, Wegener evaded causation, considering that to be premature: "It will be necessary first to exactly determine the reality and the nature of the displacements before we can hope to discover their causes." On the first page of the third instalment, there is a remarkable statement that reads as follows in translation:

> . . . the Mid-Atlantic ridge [should be regarded as] the zone in which the floor of the Atlantic, as it keeps spreading, is continuously tearing open and making space for fresh, relatively fluid, and hot sima [rock from the lower layer of the crust] rising from the depths.[9]

This hypothesis had two serious flaws. Firstly, his continents breezed across the ocean crust like icebreakers, powered in some vague manner by centrifugal and tidal forces. No geophysicist could be expected to swallow that idea. Secondly, there were errors in his own data that caused him to overestimate the rate of movement: his 2.5 metres a year is a hundred times faster than the measured rate at which North America and Europe are separating. However, initially his papers had an insignificant impact: he only published in minor journals, and there was a major war brewing.

⊙ SEEING OFF OPPONENTS, SEEKING SCARCE SUPPORTERS

When the Kaiser's army invaded neutral Belgium on August 4, 1914, Wegener was immediately conscripted as an officer in the infantry reserve. He engaged in the brutally fierce fighting in Flanders during the last two weeks of October 1914. On the opposite side, young Georges Lemaître (1894–1966), the future cosmologist, was fighting in the Belgian infantry to preserve a little corner of Belgium as unoccupied territory. Wegener and Lemaître served their respective countries for the duration of World War I.[10] Wegener was wounded twice in this time, the second time seriously with a bullet lodged in his neck that required a lengthy period of convalescence in Marburg. He thus was excused from further action on the Western Front and instead assigned to the army weather service. Although this required considerable travel to weather stations in battle zones – Germany, the Western Front, the Balkans and the Baltic – he finally had some time off duty to think.

He had published 20 papers on climatology and meteorology before the war ended, and while convalescing, he reworked and expanded his papers published in 1912 into a 94-page booklet, *The Origin of Continents and Oceans* (Figure 8.1).[11] Much new detail was added on geology when Wegener met by chance Hans Cloos (1885–1951), a young structural geologist who had arrived in Marburg in 1914. Wegener asked for his assistance, and they began a loose collaboration. In 1947, Cloos remembered:

> I liked the man very much, even though I was skeptical of his
> ideas. . . . It placed an easily comprehensible, tremendously exciting
> structure of ideas upon a solid foundation. It released the continents
> from the Earth's core and transformed them into icebergs of gneiss
> [granite] on a sea of basalt. It let them float and drift, break apart and
> converge. Where they broke away, cracks, rifts, trenches remain;
> where they collided, ranges of folded mountains appear.[12]

The immediate impact of this book was imperceptible on its publication in 1915; wartime restrictions meant it was only available in Germany. At the end of the war, Wegener succeeded his father-in-law as head of the German Naval Observatory in Hamburg, where he became highly productive. A thoroughly revised second edition of his book appeared in 1920, and this edition includes Wegener's iconic paleogeographic maps of the continents on accurate stereographic projections. The union of the continents in former times obviously gave an easy explanation for the distribution of life past and present. Wegener's reconstruction even showed that the main Carboniferous coalfields of the world were formed in the tropics. A third edition thundered off the press in 1922, and this time there followed translations into English, French, Spanish, Swedish and Russian. This set off a delayed reaction among geologists, geographers and geophysicists, whose fury sparked a firestorm, as well as a cottage industry among

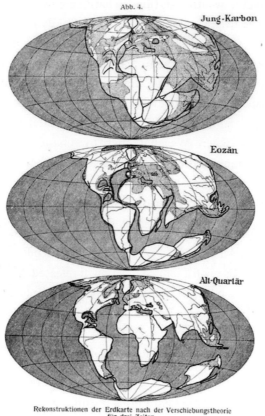

FIGURE 8.1 Alfred Wegener's paleogeographic reconstructions of the world into three periods (late Carboniferous, Eocene and older Quaternary) according to the theory of continental drift (from Wegener, 1929, fig. 4). The upper map shows the supercontinent of Pangaea.

historians who have tried to make sense of Wegener's critics. Geologists became personal in their blistering attacks on the outsider – "he's a mere meteorologist, for goodness sake" – who had no credentials as a geologist. Physicists were unmoved because Wegener did not offer a believable driving mechanism. In Cambridge, for example, Sir Harold Jeffreys (1891–1989) used elementary calculations to show that Wegener's postulate that tidal forces in Earth's crust produced the movement of the continents was impossible because the crust is too strong, otherwise mountain ranges would crumble under their own weight. In a similar fashion, he showed that the tidal force to move a continent was so great that it would stop Earth rotating within a year!

Two notable supporters of Wegener were Alexander du Toit (1878–1948) in South Africa and Arthur Holmes in Edinburgh. Du Toit was South Africa's most important field geologist, with observational and synthesis skills of the highest order. In 1923, the Carnegie Institution of Washington, DC, funded him to study the geology of Argentina, Paraguay and Brazil. He had already mapped the complete stratigraphy of 400,000 km^2 of the Karoo region of South Africa when he applied to Carnegie for support. By 1927, he was thus able compare the geological and fossil stratigraphy from the coastlines of southwestern Africa and southeastern South America, and he found that the stratigraphy was nearly identical. He became an ardent supporter of continental drift theory, publishing his findings and conclusions in *Our Wandering Continents: A Hypothesis of Continental Drifting* in 1937, seven years after the passing of Wegener. Du Toit proposed that two supercontinents had formed independently: these were Laurasia and Gondwanaland, separated by an ocean, Tethys, and the two supercontinents had collided 200–300 million years ago to form Pangaea.

By the 1920s, Holmes was in the vanguard of new thinking by British geologists.[13] As the world's expert on the age of Earth, he was famous for establishing its great antiquity: at the 1921 meeting of the British Association in Edinburgh, there was a consensus that Holmes' radiometric age of 1600 Ma was correct. Six months later, a meeting of American geologists in Philadelphia led to "a marked change of opinion in favour of the longer estimates." We can gage the change in the geoscience mindset in Britain from an airing of "the Wegener hypothesis" at the 1922 meeting of the British Association. Most speakers tentatively supported the theory, or at least said it merited further investigation. Was this polite British reserve in action or a case of selective reporting? Possibly both. More severe criticism came a few weeks later at an afternoon meeting of the Royal Geographical Society on January 22, 1923. Philip Lake, of the Sedgwick Museum, Cambridge, accused Wegener of fiddling with his jigsaw pieces and of failing to avoid bias in his data. Consider this final sentence: "... if the continental masses ever were continuous they were not fitted as Wegener has fitted them."[14] That set the scene for a lively discourse, during which one speaker remarked ironically that Wegener "is an exceedingly bad advocate for his own theory ... [whom] we have to thank ... a great deal in bringing it forward and offering himself as a target for [our] bullets." Jeffreys couldn't resist chipping in with his well-known objection: "My main complaint against the theory is that the physical causes that Wegener offers for the migrations of continents are ridiculously inadequate." On the other hand, the final speaker, the glaciologist and polar explorer Charles Seymour Wright (1887–1975), pointed out that Wegener's hypothesis was only ever intended as a general explanation of facts gleaned from many disciplines and therefore should not be criticized "on a few points of detail."

Support for his hypothesis remained muted, but its critics were not. Wegener did not go out of his way to answer those who challenged him. Instead, he switched his research theme to the climate of past eras, working with his father-in-law, Köppen, and a new collaborator, Milutin Milankovic (1879–1958), a Serbian mathematician and climatologist. *Climates of the Geological Past* appeared in 1924, without fanfare or controversy, yet today it is regarded as the foundation of the geological field of paleoclimatology.[15] Symposia and articles on Wegener's drift theory continued to be uniformly caustic and insulting. Opposition to drift theory was so strong in Germany that no university would appoint him to a chair, which meant personal finance became a challenging issue. Fortunately for him, the University of Graz, Austria, was more tolerant of controversy and supportive of academic freedom of speech. They appointed him to a professorship in climatology and meteorology. Henceforth, he concentrated on atmospheric research, apart from a trip in November 1926 during which he was on the defensive at a symposium on continental drift, convened by the American Association of Petroleum Geologists (AAPG). He was the sole participant who spoke in favor of his theory. Chicago professor Rollin T. Chamberlin (1881–1948) held rigidly to the permanence of the continents and was unsympathetic to the notion that they drifted. He is often quoted as saying: "If we are to believe Wegener's hypothesis, we must forget everything that has been learned in the past 70 years and start all over again." Inertia to change can seriously impede the progress of science. Three years later, the fourth and final enlarged edition of Wegener's *The Origin of Continents and Oceans* appeared.[11] The quotation below is taken from John Biram's English translation (Methuen, 1967) of it. I include it here because it elegantly expresses some of the guiding principles of the Deep Carbon Observatory, which is to combine facts from all branches of science, to facilitate networking in order to break down the silos that compartmentalize science and to encourage openness in the search for "truth."

> Scientists still do not appear to understand sufficiently that all earth sciences must contribute evidence toward unveiling the state of our planet in earlier times, and that the truth of the matter can only be reached by combing all this evidence ... It is only by combing the information furnished by all the earth sciences that we can hope to determine "truth" here, that is to say, to find the picture that sets out all the known facts in the best arrangement and that therefore has the highest degree of probability. Further, we have to be prepared always for the possibility that each new discovery, no matter what science furnishes it, may modify the conclusions we draw.

Wegener took his own advice seriously. In this edition, he suggested that a novel convection mechanism for the lateral movement of the continents, "recently proposed by Holmes," should be pursued.

FIGURE 8.2 The German Greenland expedition of 1930–1931. Wegener died in Greenland in November 1930 while returning from an expedition to supply food to a group of researchers overwintering in the middle of the ice cap.
Source: Archives Alfred Wegener Institute. Public domain

Holmes was doing precisely that, at the University of Durham, when Wegener tragically perished in November 1930 (Figure 8.2).

⊙ HOLMES WARMS TO CONTINENTAL DRIFT

The postwar financial crisis in Britain found Holmes in desperate difficulty. The economy of the UK had crashed from superpower status in 1914 to that of a dark country distinguished by its debts, depression, deflation, decline and mass unemployment thanks to a disastrous political decision taken in 1919 to tie the pound sterling to the gold standard. Holmes had no option but to throw in his lot with jobless academia. He signed on as the chief geologist for an oil exploration company operating in Burma. Sadly, through no fault of Holmes, that gamble failed due to a lack of investment. In 1924, he returned to England penniless and without a position. Fortunately for him, an opportunity became available at the University of Durham, where the dons had finally woken up to the fact that if they were to have a future, they needed to offer degrees in the sciences, in addition to the standard farc of divinity and the arts. Holmes was appointed reader in geology and was then quickly upgraded in 1925 to a professorship. There were no more than two or three honors undergraduates admitted each year, and since he declined to be dragged into university politics, he had plenty of time for writing and research. In this creative period, he gratefully drew on the bank of time.

The fundamental research that Holmes had carried out on geochronology had broadened his interests in the thermal history of Earth, and he was not the only one who was thinking about radioactivity as a possible agent of drift. For example, John Joly (1857–1933), an eminent professor of geology in Dublin, had worked with Ernest Rutherford on applying the radioactive decay in minerals to geochronology. In the geosciences, Joly correctly explained the microscopic pleochroic haloes in minerals such as biotite as a striking manifestation of radiation damage in rocks.[16] He read a paper at the 1926 AAPG symposium, where he offered the drift opponents an alternative mechanism. He knew well enough that the seafloor is normally too rigid to permit continents to sail through it, but he speculated that the seafloor might soften on account of the thermal energy being released by radioactive decay. Once superheated by convective currents, the floor would buckle and fracture, and under those circumstances perhaps the continents experienced lateral motion. This imaginative hypothesis was not taken up by other researchers, apart from Holmes, for whom it sparked an interest in continental drift.[17]

Holmes' professional training enabled him to take an unbiased look at drift theory, in marked contrast to the petroleum geologists in the USA. Like Joly, he had an expert's knowledge of radioactive processes in Earth's interior: his first three papers on the thermal history of a contracting Earth appeared in 1915 and 1916. However, by 1925, he was ready to retract models of a cooling and contracting planet, because gravitational contraction would not release anything like the abundant heat needed for magma generation. Only radioactive decay could do so. It was time for Holmes to renounce his assumption that Earth had been cooling down throughout its enormous lifetime, and so, from that point on, he regarded radioactivity as shaping the contours of Earth's surface.

By 1927, Holmes was freely attacking the first edition of Jeffreys' *The Earth*, the great textbook that would influence countless geophysicists for more than half a century.[18] Holmes argued that the cooling and contracting Earth concept had failed to account for magma generation, mountain-building processes and geosyncline formation. Jeffreys was evidently most put out by these criticisms: when he responded in *Geological Magazine*, he complained, "It seems to be curiously difficult to obtain fair treatment for the theory of a continuously cooling earth and its consequence, the thermal contraction theory of mountain formation."[19] Following which, he laid out the shortcomings, as he saw them, in Holmes' approach to geophysics. This somewhat schoolmasterly "Holmes must try harder" review from an alumnus of the college that became the University of Durham triggered an immediate response from the professor in charge of geology at Durham: Holmes' thoughts turned again to Earth history and radioactive heating.[20]

On January 12, 1928, Holmes announced his latest ideas on radioactivity and continental drift to the Glasgow Geological Society: he set out how convection currents in the mantle could lead to lateral flow and could thus be the long-sought driving mechanism of drift. Holmes regarded Jeffreys' objections to the absence of a driving mechanism as an inverse problem in geophysics: he thought, let us admit that continents have drifted and then see how it could work. Holmes adopted a model of the mantle that was highly viscous and heated differentially (not at the same rate throughout) by radioactivity, in which case there would be sheet-like convection upward over the location of greatest heat output, and this would drag the underside of the continental floor in two opposite directions. The return downward convection current would be just beyond the edge of the continent, which would be dragged along and down. If this sounds familiar, that's because it is! With hindsight, we can make a connection between the mechanism proposed by Holmes and the concepts of

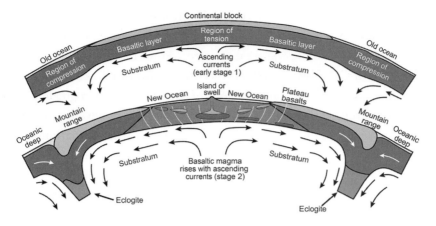

FIGURE 8.3 The final 1948 version of Holmes' model, first published in 1931. Holmes captioned it thus: "Diagrams to illustrate a purely hypothetical mechanism for 'engineering' continental drift." In the upper diagram, the subcrustal currents are in the early part of the convection cycle. In the lower part, "the currents have become sufficiently vigorous to drag the two halves of the original continent apart."
Source: Penny Wieser

subduction, mantle plumes and so on (Figure 8.3). But that's racing ahead.

By the time Holmes published an expanded version of his convection hypothesis in 1931, he felt that he had finally cracked the puzzle of continental movement. The caption that accompanied the 1931 version of the image, showing the convection currents ascending and descending, has the intriguing suggestion that: "The formation of a new ocean floor would involve the discharge of a great deal of excess heat." This hypothesis, of a spreading ocean floor as the cause of continental drift, failed to move any opponents. Jeffreys had this to say at a meeting of the British Association in 1931:

> I have examined Professor Holmes' theory of subcrustal currents to some extent, and have not found any test that appears to be decisive for or against. So far as I can see there is nothing inherently impossible in it, but the association of the conditions that would be required to make it work would be rather in the nature of a fluke.[21]

During the 1930s, there was very little interest in Holmes' hypothesis: it did not have the status of an earthshaking idea and was deeply unfashionable among the establishment. From his tranquil chamber in St. John's College, Professor Sir Harold Jeffreys FRS continued to chunter away quietly about the folly of the drifters. Although Holmes had constructed his theory carefully in terms of the physics and the consequences of the model, it had few followers because of its appeal to radioactivity as the driver of convection.[22]

The early 1940s were the low-water mark for drift theory. Once again, Holmes had time on his hands: there were few students of geology to teach, and he would often stand night watch for German incendiary bombs at the Durham laboratories. He reacted positively when a London publisher suggested that he should write an up-to-date account of physical geology. Thus began his second period of great productivity in writing his *Principles of Physical Geology*. This copiously illustrated textbook covered the entire range of Earth processes as they were understood at that time. He included current views on topics that remained controversial: the work of Wegener is in the final chapter, along with an outline of Holmes' convection model. He continued to make modest claims: describing the model as "purely speculative," and stating that "many serious difficulties still remain unsolved." The masterpiece was reprinted annually from publication in 1944 to the early 1960s.

When Sir Kingsley Dunham (1910–2001) wrote the Royal Society's biographical memoir of Arthur Holmes, he paid scant attention to the work on "Earth movements," simply noting what Holmes had said at the AAPG meeting of 1926 and the Glasgow lecture in 1929. He continued:

> Subsequently continental drift encountered severe opposition from geophysicists and geologists alike, but in the past decade Holmes has had the satisfaction of seeing the hypothesis revived with strong support from palaeomagnetic evidence, and from the remarkable "fit" of the continental shelves on opposing sides of the Atlantic where plotted in spherical projection.

Here and in the previous chapter we have made a journey lasting a century. This began with Kelvin's investigations of Earth's internal heat energy, and now we have arrived at the moment when the hundred-year deadlock between the drifters and the fixists is about to end. In our saga, a new dawn breaks on the horizon of the fields of geophysics, geochemistry and geobiology. In the next half-century, this geoscience enlightenment period set the scene for many new *interdisciplinary* lines of inquiry, one of them being deep carbon science.

☉ WHY DID IT ALL TAKE SO LONG AND WHAT WAS LEARNED?

Scholars in the history and philosophy of science have documented and debated in considerable detail the question as to why it took a century to understand the complexities of Earth's interior. The century of controversy confronted three vexing problems: establishing the age of Earth; understanding the limitations of the laws of thermodynamics when applied to Earth's interior; and accounting for how continents could move through the ocean floor. Each of these research programs became an exercise in *physics* for the theorists and a search for *data* by the geologists. Given the paucity of hard facts, it is scarcely surprising that geology's Great Debates frequently crystallized into matters of "belief." When new facts challenged the consensus, models would undergo contingent adjustment to accord with reality. This methodology was by no means unique to geoscience. In astronomy and cosmology, the procedure of adjusting models dates back to antiquity. In the ancient natural sciences, the procedure acquired the slogan "saving the appearances," an exercise that lasted for two millennia, until the Copernican Revolution. In the history of geoscience, the belief issues had variously included Darwinian evolution, radioactivity as the source of internal heating and convection in the mantle. Belief is a poor counselor in academic debates if it substitutes for the reality discoverable in experiments or by observation. And that strikes at the heart of the problem. Both the field geologists and the theoretical physicists were trying to fathom the inaccessible interior of Earth by puzzling over the rock formations, rifts and ridges of the surface. Technical improvements in geology were delayed for several decades. The essential developments in instrumentation and techniques date only from the early twentieth century in the case of seismology, and later still in fields such as oceanography.

The debate about the geological and geophysical properties of the interior was also hindered by a reluctance to set aside long-cherished research programs. In the absence of compelling evidence for a novel hypothesis, there is an understandable reluctance for researchers to become sidetracked from the general consensus. As we have seen, American geologists were united in rejecting continental drift, largely because there were insufficient data to shift their perception of the hypothesis. The fixists stood their ground by accusing the mobilists of being too selective with their data. Any early-career academics who might have been inclined to support drift were cautious, realizing that it was just too risky to speak out in the absence of evidence.

Kelvin and Jeffreys were among the outstanding theoretical physicists of their generation, with a strong belief in the *laws* of physics. There is a legend that in 1900 Kelvin had said, "There is nothing new to be discovered in physics now. All that remains is more and more precise measurement." At the time of Kelvin's passing, radioactive heating had been suggested by Pierre Curie, Jeffreys was embarking on his ordinary BSc degree in Durham, Holmes was starting his physics BSc in London and the meteorologist Wegener was on his first Greenland expedition. All of them would have been impressed by Kelvin's immense reputation in *classical* physics. The Royal Society's biographical memoir of Jeffreys, written by Alan Cook (1922–2004), begins:

> Harold Jeffreys stood out among the small group of pioneers who
> developed the physical study of the Earth from its primitive
> condition at the beginning of the 20th century to its state at the

launch of the first Sputnik. He, above all, applied classical
mechanics to investigate the interior of the Earth.

Disharmony between geophysical and geological communities continued for so long because of the persistent absence of compelling
evidence, not simply from the field, but also from the interior.
Holmes wrote as much several times during his career. As we shall
see in the next chapter, the great leap forward began when scientists
in Britain and America returned from war work in applied disciplines
such as nuclear physics, radio science, radar, sonar and so on and
repopulated university laboratories. While working in the military–
industrial complex, they had learned the importance of international
teamwork, well-defined missions and state-of-the-art instrumentation, as well as data acquisition and analysis.

REFERENCES

1. Frankel, H. Hess papers, Princeton University. Letter to Vening Meinesz dated
   6 July 1959. Harry Hess develops seafloor spreading. In *The Continental Drift
   Controversy* (Cambridge University Press, 2012), pp. 198–279.
2. Wegener, A. Die Einstehung der Kontinente. *Geologische Rundschau* **3**,
   276–292 (1912).
3. Powell, J. L. Book review of Alfred Wegener: Science, Exploration, and the
   Theory of Continental Drift by Mott T. Greene. *Isis* **107**, 871–872 (2016).
4. Romm, J. A new forerunner for continental drift. *Nature* **367**, 407–408 (1994).
5. Rupke, N. A. Continental drift before 1900. *Nature* **227**, 349–350 (1970).
6. Greene, M. T. *Alfred Wegener: Science, Exploration, and the Theory of
   Continental Drift* (John Hopkins University Press, 2015).
7. Hoffman, P. F. The tooth of time: Alfred Wegener. *Geoscience Canada* **39**,
   (2012).
8. Jacoby, W. R. Translation of *Die Einstehung der Kontinente* by Alfred Wegener,
   part 1 of *Petermanns Geographische Mitteilungen*, volume 58, 1912. *Journal of
   Geodynamics* **32**, 29–63 (2001).
9. Jacoby, W. R. Modern concepts of Earth dynamics anticipated by Alfred
   Wegener in 1912. *Geology* **9**, 25–27 (1981).
10. Mitton, S. The expanding universe of Georges Lemaître. *Astronomy &
    Geophysics* **58**, 28–31 (2017).

11. Wegener, A. *Die Entstehung der Kontinente und Ozeane* (Vieweg & Sohn, 1915).
12. Cloos, H. *Gespräch mit der Erde* (Verlag Piper & Sohn, 1947).
13. Lewis, C. *The Dating Game: One Man's Search for the Age of the Earth* (Cambridge University Press, 2002).
14. Lake, P. Wegener's hypothesis of continental drift. *Geographical Journal* **61**, 179–187 (1923).
15. Köppen, W. and Wegener, A. Die klimate der geologischen vorzeit. *Quarterly Journal of the Royal Meteorological Society* **51**, 287–289 (1925).
16. John Joly, 1857–1933. *Biographical Memoirs of Fellows of the Royal Society* **1**, 258–286 (1934).
17. Frankel, H. Arthur Holmes and continental drift. *British Journal for the History of Science* **11**, 130–150 (1978).
18. Jeffreys, H. *The Earth: Its Origin, History and Physical Condition* (Cambridge University Press, 1924).
19. Jeffreys, H. The Earth's thermal history and some related problems. *Geological Magazine* **63**, 516–525 (1926).
20. Holmes, A. Some problems of physical geology and the Earth's thermal history. *Geological Magazine* **64**, 263–278 (1927).
21. Mackinder, H. *et al.* Problems of the Earth's crust, a discussion in Section E (Geography) of the British Association on 28 September 1931 in the Hall of the Society. *Geographical Journal* **78**, 433–455 (1931).
22. Frankel, H. The career of continental drift theory. *Studies in History and Philosophy of Science Part A* **10**, 21–66 (1979).

9   The Mid-Atlantic Ridge and Rift Valley

A fundamental goal of geophysics is the construction of a comprehensive understanding of the structure and dynamics of Earth's interior, which is inaccessible to direct measurement. Therefore, indirect methods such as the interrogation of seismic data obtained at the surface are used to model the physical properties of the interior. Interpretation of data requires the researcher to solve an inverse problem. In scientific inquiry, the solution of an inverse problem requires interpretting data to discover underlying causes. When we ask, "Why is it raining today?" we are posing an inverse problem: we have the data (getting wet!) and want to know the cause (frontal system, thunderstorm, etc.). The reverse situation – posing a forward problem – is to ask, "Will it rain tomorrow?" This can be tackled by computing a forecast from an atmospheric model. This chapter outlines an important case history in which the mere acquisition of data for its own sake led to a fascinating inverse problem: Why is the longest and greatest mountain range on Earth in the middle of the oceans?

Surveying and mapping of the ocean floors for commercial interests and warfare at sea led to two very great discoveries: the system of mid-ocean ridges and their central rift valleys; and the deep ocean trenches. There is a fascinating backstory to recount as well. As geophysics developed into an eminently practical science, these scientists profited from varied connections to the political classes, the military and commercial enterprises. This is the period when "big science" and "big data" greatly enlarged the scope of geoscience: when fieldwork migrated from land to the oceans and dedicated research ships first plumbed the depths.

So, we now take the plunge and dive into oceanography, a rich and diverse interdisciplinary subject concerned with all aspects of the world's seas and oceans. About two-thirds of Earth's surface lies under the oceans, and many of the geological processes occurring on land have links with the dynamic ocean floor. Oceanography intertwines several disciplines: biology and ecology, physics and chemistry, Earth history and geology. As a global discipline, oceanography contributes to our narrative on deep carbon science on account of the contribution oceanographers made to understanding plate tectonics. In later chapters, we shall see how ocean drilling programs and the discovery of hydrothermal vents contributed to our understanding of biogeochemistry and the origin of life.

A convenient point of embarkation for the next stage of our journey is the publication in 1855 of the first comprehensive textbook on oceanography: *The Physical Geography of the Sea*, written by Matthew Fontaine Maury (1806–1873), who was then the superintendent of the US Naval Observatory (USNO). Back in 1842, as a young lieutenant, Maury had been in charge of vast collections of charts and maritime logbooks at the US Navy Office in Washington, DC. Perspicacious scrutiny of the records (field notes!) of countless master mariners enabled him to produce reliable ocean current and wind charts for use by navigators who were planning lengthy ocean voyages. Following his outstanding success in combing through logbooks to crowdsource and interpret physical data, Maury became convinced that knowledge of the oceans could only be obtained by international cooperation and the sharing of data.

When Maury made that suggestion, scientific investigation of the seas and oceans had been limited to navigational aspects, such as depth soundings in coastal waters, and the knowledge of tides and currents. Edmond Halley directed the first sea journey made for purely scientific purposes. This was his expedition of 1698 for making observations of magnetic declination on board the Royal Navy's six-gun, square-masted, pink HMS *Paramour*. Halley's data had an immediate application in a military context: in 1702, the English were at war with France and had deployed two squadrons of 50- and 60-gun frigates to defend the North Sea and the Western Approaches.

The Royal Navy then waited for a century and a half before engaging once more in oceanographic research. In 1868, Charles Wyville Thomson (1830–1882), a natural historian with a keen interest in the biological conditions of the deep seas, persuaded the Royal Navy to convert an old paddle steamer (the sixth HMS *Lightning*, launched 1823) for use as an oceanographic survey vessel. He used her to conduct deep-sea dredging experiments and hydrographic measurements off the coast of Norway. He found that animal life existed down to depths of 650 fathoms (1200 metres). The fascinating results on life in the deep sea stimulated Thomson to persuade the Royal Society to seek government support for a prolonged voyage of scientific exploration. Political agreement came remarkably quickly. George Goschen MP (1831–1907), the First Lord of the Admiralty, signed off the use of HMS *Challenger* together with its complement of 200 officers and ratings for a 4-year global expedition. The substantial costs were underwritten by the Chancellor of the Exchequer, Robert Lowe MP (1811–1892), who was a passionate advocate of education reform. The British government and Queen Victoria's Royal Navy thus became fully involved in publicly supporting a huge scientific project without parallel in nineteenth-century Europe. For the Admiralty, an important outcome of this expenditure was to be its survey of the depth of the ocean floor. The Admiralty's motive was to enable the laying of submarine telegraph cables for transmitting orders to warships in the far-flung realms of the British Empire. Above all, their support for oceanographic science was about sending a symbolic message to demonstrate Britannia's ability "to rule the waves."

⊙ THE *CHALLENGER* EXPEDITION, 1872–1876

To convert the three-masted naval warship into a research vessel, all but two of her 17 guns were removed. In their place came additional cabins for the six scientific staff recruited by Professor Thomson. Two

FIGURE 9.1 The laboratory on the former gun deck of *Challenger*.
Illumination is by candles and oil lamps. There are three microscopes for
the examination of samples and a quill pen for writing field notes.

laboratories and workrooms provided ample storage space for the
sample collections (Figure 9.1). A special dredging and sounding platform was also provided on deck. She steamed out of Portsmouth just
after noon on December 21, 1872, coincidentally at precisely the same
moment as the winter solstice. When she called at Lisbon for provi-
sions and 150 tons of coal, the King and Queen of Portugal came on
board in the afternoon to view the scientific apparatus. In the evening,
the Captain and the Professor were invited ashore to dine with the
King.[1]

At each sampling station, Thomson stood on deck supervising a
standard sequence of observations: total depth of the water, water
temperature at various depths, weather conditions, as well as the
direction and speed of the surface current, and currents at different
depths if possible. During the sampling routine, the naval crew used
the steamship engine for maneuvering and steadying the ship for the
scientists and their equipment. The deckhands dredged or trawled

FIGURE 9.2 Demanding and dangerous work: obtaining the samples at each station.

samples of sediment that included fauna from the seabed and they dipped for samples of water and marine life at intermediate levels. Soundings were made by lowering a hemp rope with a 50-kg weight over the side, to which were attached thermometers, water bottles and sinkers (Figure 9.2). This process could take a whole day at deep locations. The repetitive procedures took place at 362 sampling stations, and it soon became a drudge for the 200 naval ratings. During deep-water dredging, the officer of the watch had to remain on his feet as overseer for up for 12 hours, while the ratings did much heavy lifting of gear. When the chug-chug of the donkey engine pulling the ropes presaged the imminent arrival on board of further samples from the watery deep, the scientists were bright-eyed in anticipation, but the deckhands were physically exhausted and bored by the monotony of ocean sailing. No surprise, then, that when in port in Australia several crew members jumped ship. During *Challenger*'s circumnavigation of the globe, a further seven crew died of accidents or illnesses and were buried at sea. She sailed nearly 70,000 nautical miles (127,580 kilometres) while carrying out scientific surveying and exploration.

When *Challenger* arrived at home on May 24, 1876, she carried a bounteous hoard of research material for study and classification, as well as tens of thousands of photographs and drawings. It would take 100 scientists about two decades to write up the discoveries made with these samples. Queen Victoria rewarded Thomson with a knighthood for his contributions to ocean science. Marine biology of the deep was the hugely successful scientific focus of the expedition: more than 4700 new species of marine animals were discovered. The life deep on the seafloor astonished the biologists, who had assumed the abyss would be too inhospitable for life. Research on marine deep life, a major strand of the Deep Carbon Observatory pathway, began with *Challenger*. The discovery of the richness and variety of creatures in the deep ocean fascinated the general public: it was the biggest science news story since Darwin had stepped off HMS *Beagle* 40 year earlier. Arabella Buckley (1840–1929), who had been acquainted with both Lyell and Darwin, authored popular science books for children. As a vivid writer, she knew how to captivate a young readership:

> ... these curious fish, living in eternal darkness ... many of them with their own lights ... What slaughter and hunting there is among them! for they all eat each other ... Strange monsters are all these deep-sea fish, some of them living as much as 16,000 feet under the surface of the sea.[2]

The government was well aware of the need to make the scientific discoveries widely available: they appointed a Commission to arrange the collections and ensure publication of the results, all at public expense. Sir Charles Thomson was put in charge of the project, despite being in poor health, the long years at sea having weakened his constitution. The dull administrative work of commissioning detailed illustrations for 50 volumes of scientific information became too stressful for him: in the summer of 1879, he stopped giving lectures. In 1881, he passed the immense editorial task over to John Murray (1841–1914), who had been responsible on the voyage for investigation of the natural history of the seafloor, as well as the collection of hundreds of tons of sediment from the abyssal plains. Murray worked intensively until 1895 on the 50 volumes of the *Report on the Scientific Results of the Voyage of* HMS *Challenger*. When the government's grant expired, he paid all of the remaining costs from his own pocket. The report was the work of numerous experts from many countries and ran to 29,500 pages. Murray summed up the significance of what had been achieved by calling it "the greatest advance in the knowledge of our planet since the celebrated discoveries of the fifteenth and sixteenth centuries." By the latter he meant the Copernican Revolution.

⊙ FINDING THE MID-ATLANTIC RIDGE AND THE MARIANA TRENCH

The two great scientific achievements of the expedition were mapping the huge extent of the Mid-Atlantic Ridge and the finding of the deepest parts of the ocean in the western North Pacific. During the first year of the expedition, *Challenger* crossed the Atlantic Ocean four times, and when homeward bound she sailed in the mid-ocean from Tristan da Cunha to the Cape Verde Islands. These soundings of the Atlantic were urgently needed for mapping the contours of the seafloor so that expensive communication cables could be laid down without too much risk of them being lost or broken.

On September 30, 1852, USS *Dolphin*, a 10-gun brig, put to sea from the New York Navy Yard on a special cruise in support of Lieutenant Maury's investigations. The warship made soundings in the Mid-Atlantic near the Azores, where the ocean was found to be surprisingly shallow. Maury made a contour map of these soundings that revealed a continuous shallow zone north of the Azores, which he termed the Azores Middle Ground: this was the first name for the Mid-Atlantic Ridge. The two *Challenger* crossings made in 1873 confirmed Maury's findings and widened the field of inquiry: What was the extent of this rise in the seafloor? The homeward run from Tristan to Cape Verde in 1876 provided the answer to that question: soundings along the Ridge revealed its continuity and vast length.

On March 23, 1875, *Challenger* sounded a staggering 8200 metres (4475 fathoms) at station 225 in the Pacific. That great depth astonished the scientists, even though they had already sounded 3875 fathoms in 1873 between the Caribbean Islands and Bermuda. They made two more soundings to ensure their accuracy, but they had not broken the record for depth: by now *Challenger* was not alone in surveying the Pacific. The honor of plumbing the greatest depth was held by USS *Tuscarora*, which had sounded 4655 fathoms when surveying the cable route from California to Japan in 1873–1874. The Germans launched a round-the-world cruise of their own using SS *Gazelle*, which was in the Pacific contemporaneously with *Challenger*, and efforts were made to coordinate the surveys.[3] The results from these voyages showed that the Pacific was different from other oceans. It was uniformly deeper, with much of the seabed lying at depths greater than 3600 metres (2000 fathoms). However, these sounding had not yet revealed the existence of the famous Mariana Trench: that would take another 30 years.

In 1877, the accomplished cartographer Augustus Petermann (1822–1878) used *Challenger's* data to publish the first bathymetric chart of the Pacific Ocean. He named the submarine valley probed by *Challenger* as the Challenger Deep. John Murray's maps were published later, in 1895 and 1899, and they depict it as a small, circular area enclosed by the 4000-fathom isobath. This is the deepest known point in Earth's seabed hydrosphere (10,994 metres). It lies at the southern end of the Mariana Trench, the first hint of which came in November 1899 when USS *Nero* sounded 5269 fathoms (9636 metres) in the course of surveying the telegraph cable route between the USA and the Philippines. The northern end of the Trench was sounded out in 1902 by the cable-laying ship *Colonia*. This crescent-shaped scar in the ocean floor is about 2550 kilometres long and its average width is 70 kilometres. The first outline map of the Trench was published in 1907, 30 years after the initial sounding by *Challenger*.

The discoveries of the Mid-Atlantic Ridge and the Mariana Trench are reminders of the extreme difficulties and risks that the pioneer investigators faced in sounding out the ocean basins and subsequently extracting the physical geology of the seafloor from their data. In an era of heroic seafaring, they obtained sparse soundings with imaginative but inadequate equipment. And yet it was they who first discovered the ridges, rifts and basins of the world's oceans. They mapped the continental shelves and their slopes, the abyssal plains and small features such as canyons and seamounts. Making sense of how and why these features had arisen had to wait until the next big breakthrough in experimental technique, however: the echo sounder.

⊙ THE POSTWAR BONANZA FOR GEOSCIENCES

From 1945 to the mid-1960s, there were stupendous advances – as well as cliff-edge transitions – in the physical sciences. When young research scientists in America and Britain returned from war work to their universities and research institutions, many of them had experienced new ways of doing science selflessly, as members of interdisciplinary teams and subject to strong leadership as well as urgent deadlines. Group structures now brought together experimenters and the theorists, as well as electronics specialists and instrument designers, all working in cooperation rather than competition. These early-career scientists of the mid-1940s had skipped doctoral research and thesis writing, having served their academic apprenticeships on defense projects for their countries. Well before Hiroshima, World War II had demonstrated the power of science and technology in military applications such as code-breaking using computers and the development of sonar and radar for remote sensing. In Britain, the jolly term "boffin" entered the language as a slang term for the radio engineers and scientists contributing to the defense efforts. Geophysical research expanded rapidly in areas such as atmospheric and ionospheric physics, seismology, oceanography and geomagnetism. All of these fields were to benefit from the new global approach to the acquisition of data.[4]

On April 5, 1950, rocket scientist James van Allen (1914–2006) hosted a small dinner for atmospheric physicists, during which they discussed the need for a truly comprehensive approach to research in geophysics. In the following months, this suggestion developed into an ambitious proposal to organize the International Geophysical Year (IGY) 1957–1958 to coordinate a mass attack on problems concerning the physical structure of Earth and its atmosphere. Tens of thousands of physical scientists from 67 countries participated its programs to improve knowledge of the atmosphere, continents and oceans. By the completion date of the IGY in 1958, the power of international scientific cooperation in geoscience research programs had been convincingly demonstrated. A global approach to data acquisition and interpretation had been a key element in the success of the IGY, and this was a foundational principle of the Deep Carbon Observatory.

The history of the "plate tectonic *revolution*" is a tale of rapid *evolution* in the philosophy of geoscience. More than two dozen scientists did the key work that created the theory of plate tectonics. It took two decades for the consensus to emerge, as scientific knowledge matured and stabilized. Several of the participants probably did not anticipate that when they began their work it would contribute to an evolution of theory as great as that of Copernicus. Plate tectonics revealed Earth's interior to be a stupendous heat engine, forcefully driving convection in the mantle so powerfully that it could move the continents and bury the ocean floor beneath the continents by plate subduction, all of which put geochemistry and geophysics in the front line for understanding the deep cycles of burial, residence and eventual exhumation of the chemical elements.

⊙ GÜNTHER DIETRICH PROFILES THE RIDGE AND THE CENTRAL RIFT VALLEY

The earliest known reference to underwater sound is in Leonardo da Vinci's *Notebook* of 1490, where he writes that by listening at one end of a long tube placed in the sea "you will hear ships at a great distance." In 1859, when Maury revised *Physical Geography*, he wrote of "ingenious and beautiful contrivances for deep-sea sound-ings," and he suggested that "by exploding petards, or ringing bells in the deep sea" an echo from the bottom might be heard, from which the depth could be inferred. These trials to detect sound propagating vertically failed, probably because the listening devices were above the surface.

For the next 50 years, range measurement at sea was devoted to the horizontal detection of objects in the water. The objects of interest included icebergs, in response to the RMS *Titanic* disaster on April 15, 1912, and enemy submarines in World War I. In the interwar years, the US Navy used sonic depth finders for coastal surveys, and sound-ings proliferated. The first scientific application of deep-water echo sounding was in 1925–1927 by the German survey vessel VS *Meteor*, which was financed by the Weimar Republic to reassert Germany's scientific standing and national pride following World War I. *Meteor* logged 13 crossings of the Atlantic Ocean, which yielded profiles from the northern tropics (20°N) to as far as Antarctica (55°S). This grand reconnaissance was undertaken primarily to see if gold could profit-ably be extracted from seawater. This speculative adventure was launched because the Treaty of Versailles (1919) had required Germany to cough up 50 tonnes of gold – an impossible demand – in reparations for civilian losses. The emphatic answer to this quest for treasure from the ocean was: no! Nevertheless, there was a great bonus for geoscience in the form of 67,400 echo soundings of the rugged topography and huge extent of the Mid-Atlantic Ridge.

Günther Dietrich (1911–1972) was the scientist in charge of echo sounding during a later expedition, when *Meteor* traversed the Atlantic four times in the spring of 1938. When he wrote up the morphological findings in 1939, he published a spectacular cross-section of the crest of the Ridge (Figure 9.3), revealing a deep depres-sion in the center. He remarked that:

> ... the sea bottom in the middle of the Mid-Atlantic Ridge drops
> from a peak height of 1,930 m to 4739 m ... within a distance of

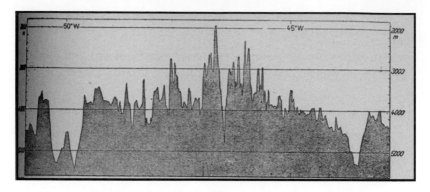

FIGURE 9.3 This profile, made from unreduced echo-sounding data from the 1938 cruise of the German survey ship *Meteor*, established oceanographer Günther Dietrich (1911–1972) as the first to note the deep median valley on the Mid-Atlantic Ridge.

Source: Some Morphological Results of the Cruise of the Meteor, by G. Dietrich, Berlin. *International Hydrographical Review.* NOAA Central Library Historical Collection

21 kilometres ... Such striking depressions in the Ridge occur also in the other five *Meteor* profiles.[5]

Dietrich did not know how to interpret these depressions because the profiles were quite widely separated along the ridge. Not surprisingly, because World War II caused an abrupt break in scientific oceanography, Dietrich's short paper in German with its big ideas was ignored for nearly 10 years, until the subject was revived in the USA. It was only much later that Dietrich could look back to claim priority in being the first to have recorded the discovery of the median rifting of the Mid-Atlantic Ridge.

⊙ HARRY HESS: "DROWNED ISLANDS OF THE PACIFIC BASIN"

On December 8, 1941, the day after the Japanese attack on Pearl Harbor, Harry Hess (1906–1969), professor of geology at Princeton University, departed on the 07:42 train from Princeton Junction for New York's Penn Station to sign up for active duty as a naval reserve officer. Hess already had plenty of experience of fieldwork at sea from running geophysics experiments with the Navy. In 1932, while a doctoral student at Princeton, he had assisted in gravity measurements on board the submarine USS *Barracuda*, cruising in the West Indies. Felix Vening Meinesz (1887–1966), a Dutch geophysicist who pioneered the study of gravity anomalies at sea, directed the operations and provided a precision gravimeter that he had invented. Maurice Ewing (1906–1974), a geophysicist who was a mere 12 days older than Hess, also participated. After the cruise, when it came to writing up the results, the two men had a falling out over the interpretation of the data. Ewing was of a judicial mind, and he would not commit to making an unsustainable claim, whereas Hess wanted to rush into print with novel hypotheses and explanations. He did just that, dashing off what became a major paper on island arcs and their origin.[6] In 1937, Ewing and Hess published separate papers on the geophysical and geological data they obtained while on USS *Barracuda*.[6,7] The edginess in their professional relationship intensified later when Ewing founded the Lamont Geological Observatory in 1949.

In 1942, Hess's first assignment was ashore in New York, where he was tasked with predicting the probable of activities of U-boats. When he then requested to go to sea, he was first given command of a decoy vessel, USS *Big Horn*, so that he could test the effectiveness of his detection technique in the field. After that, he was assigned as captain to an assault transport ship, USS *Cape Johnson*, in which role he took part in the Battle of Leyte Gulf (October 23–26, 1944), which some historians consider to be the largest naval battle in history.[8] By the end of the war in the Pacific, he was Commander Hess. Upon resuming his research at Princeton, he maintained an active status in the Naval Reserve with a further promotion, which meant he simultaneously held the positions of a professor of geophysics at Princeton and a Rear Admiral of the Fleet!

While captaining *Cape Johnson*, Hess marshaled his natural curiosity to excellent effect and contributed significantly to our understanding of Earth's dynamic interior. He was able to do so because his ship, like most in her class in World War II, was fitted with sounding gear. When not constrained by battle orders, Hess ran the sounding equipment continuously. Judiciously, he navigated routes that were not always in strict accordance with the course sent out by the higher authorities. His covert bathymetry greatly extended our knowledge of the topography of the floor of the Pacific Ocean. The data he curated while at war later led him to major advances in the investigation of ocean basins. He was successful because instead of conducting specific point-to-point soundings like the cable operators, he ranged more widely across the uncharted depths. For example, he noted about 100 isolated, flat-topped mountains submerged between Hawaii and the Marianas. He gave the name "guyot" to these quite extraordinary structures on the seabed to honor the first professor of geology at Princeton, Arnold Guyot (1807–1884). He poetically described them as "drowned islands of the Pacific Basin," and within a couple of decades some ten thousand were had been identified. Hess speculated that guyots were extinct volcanoes that had subsided into the ocean, where wave action eroded their peaks to flat tops. Given that the tops of these seamounts were now thousands of metres below the ocean surface, this hypothesis required an as-yet undiscovered dynamic machine to reposition them. We will come to that soon, but first we need to catch up with Ewing.

◉ MAURICE EWING: "I KEEP MY SHIP AT SEA"

Ewing was also working hard on marine geology, specifically on seismic reflection and refraction in ocean basins. He had begun this topic back in the fall of 1926 while a doctoral candidate in the department of physics at Rice University: his PhD thesis was entitled "Calculation of ray paths from seismic time-travel curves." This area of expertise became critically important in the early Cold War years, when civilian scientists in the Pentagon declared data from the entire field of oceanography a closed book because of its potential applications in submarine warfare. For example, the Navy urgently required improved maps of the seafloor to plot where submarines could travel undetected, to be aware of the seamounts and guyots that posed collision risks and to know where submarines could hide from enemy sonar. Good maps were also necessary for establishing a secret network of listening devices linked by cable to track Soviet submarines.[9]

In the late 1940s, the two major US mapping efforts were based at Woods Hole Oceanographic Institution on the east coast and at the Scripps Institution of Oceanography, La Jolla, on the west coast. The decisive push was made by a third institution, the Lamont Geological Observatory, established in 1949 by Columbia University to pursue new research opportunities in geophysics. This observatory was Ewing's brainchild, and as its first director he used his intense drive to build up Lamont as a world-class research center for geochemistry, seismology and marine geology.

In 1948, Ewing brought aboard Bruce C. Heezen (1924–1977), an outspoken graduate student who had already gained the right experience as a chief scientist at sea running a continuous echo sounder. At about the same time, Lamont hired Marie Tharp (1920–2006) as a drafter of maps. In World War II, she had been 1 of 10 female students recruited for the petroleum geology program at the University of Michigan. There, she gained her master's degree in geology, then worked for a bachelor's degree in mathematics at Tulsa. On finding that there were no jobs for women geologists in the petroleum industry (because the men had just returned from the war), she moved to New York, where Heezen recruited her as his shore-based assistant. Ewing and his graduate students took part in cruises on the Woods Hole research ship *Atlantis* (Figure 9.4). As a result of this collaboration, Lamont amassed tens of thousands of depth measurements from the North Atlantic.

Tharp's role was to transform the soundings into highly detailed seafloor profiles. In 1999, Marie recalled that 50 years earlier she was trying to make sense of "a hodgepodge of disjointed and disconnected profiles of the sections of the North Atlantic floor . . . the ship's tracks looked like a spider's web." One day, when piecing together the

FIGURE 9.4 RV *Atlantis*, a two-masted, steel-hulled ketch, operated by Woods Hole Oceanographic Institution from 1930 to 1933. In 1947, her research scientists identified the first abyssal plain, south of Newfoundland. It comprises nearly 1 million square kilometres, varying in depth from 4600 to 6100 metres.
Source: Woods Hole Oceanographic Institution

profiles into the correct order, she was struck by similarities of shape for half a dozen profiles across of the Ridge shown in Figure 9.5. When she checked this carefully, only one feature persisted from profile to profile: a V-shaped notch at the center of the profiles was a good match across all profiles. Her immediate thought was "it might be a rift valley that cuts into the ridge at its crest and continues all along its axis."[10] When she showed this to Bruce Heezen, he said, "It cannot be. It looks too much like continental drift." In the early 1950s, continental drift was still regarded in the USA as a form of scientific heresy. Heezen and Ewing strongly believed that the major geophysical features in the oceans were consequences of the expansion of Earth. Much later (in 1974), Bruce said in an interview on the events of 1952 that:

> Marie's job for me was to decide what a structure was – whether a rise in the echo soundings represented a hill or something longer like a ridge – and to map it. In three of the transatlantic profiles she noticed an unmistakeable notch in the Mid-Atlantic Ridge, and she decided that they were a continuous rift valley and told me. I discounted it as girl talk and didn't believe it for a year.[11]

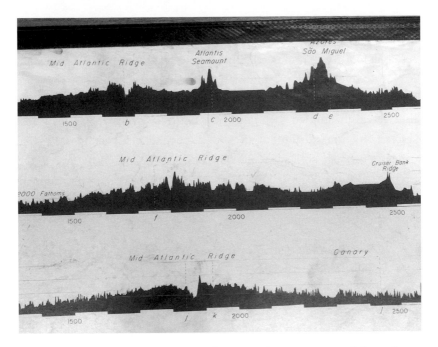

FIGURE 9.5 Marie Tharp's profiles, with her annotations of the Mid-Atlantic Ridge and its central valley. This is her handiwork pasted onto a display board, similar to a modern poster paper, which she used in seminars.
Source: US Library of Congress, Washington, DC. Simon Mitton

This was about the time that Bruce and Marie formed a lifelong relationship. Despite their close friendship, Tharp's name isn't on any of the key papers that Heezen and others published in 1959–1963.

In 1953, the US Navy unexpectedly discontinued its arrangement to supply a ship to Lamont. Like all directors of oceanographic institutions, Ewing needed access to the deep oceans, where field research was far more expensive than short excursions in coastal waters. Professionals in atmospheric and space science in this era were strongly supported by the military, as mentioned at the beginning of this chapter. But that was not so for marine research. So, in 1953, Columbia University first leased and later purchased an antiquated sailing ship, the *Vema*. She was a three-masted, iron-hulled racing yacht that had lain abandoned on Staten Island for several years until being salvaged as a charter vessel. Ewing converted her into a research facility that majestically circled the globe for 320 days a year, exploring every ocean. Lamont researchers were able to investigate broad swathes of the deep oceans as never before. *Vema* became the first research ship to sail a million nautical miles in the cause of science; by her retirement in 1981, she had clocked up 1,225,000 miles. A British colleague, Edward Bullard (1907–1980), once asked Ewing, "Where do you keep your ship?" In response, he quipped, "I keep my ship at sea." On Ewing's passing in 1974, Bullard penned this evocative epitaph:

> It was he, more than any other man, who provided the fuel for the revolution in earth science.[9]

This surely justifies counting Maurice Ewing as a pioneer of deep carbon science.

☉ THE HEEZEN–THARP MAP, 1957

The purchase of *Vema* put Heezen and Tharp in an enviably advantageous situation for drafting new maps of the Atlantic and other world oceans. Ewing was ruthlessly determined to maximize the scientific output of his research ship. To that end, he commanded his scientists on oceanic cruises to run multiple instruments and to plumb high-resolution depth profiles at every opportunity. As a result, Lamont led the world as a tidal wave of geoscience data flooded into the observatory. Throughout the IGY, the schooner *Vema* was at sea almost continuously, surveying the ocean basins and their ridges. It was a quest that would reveal the rift and ridge system girdling the world. However, publication of these results was not straightforward.

For reasons of national security, the Pentagon classified a broad range of geoscience data in 1952. Two major weapons projects of the Cold War – ballistic missiles and anti-submarine warfare – required precise data on gravitational measurements and accurate seafloor contours. To prevent breaches of security, the Pentagon prohibited

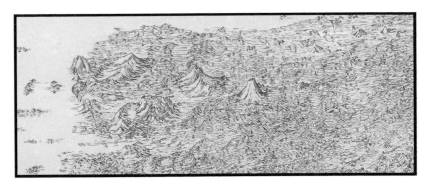

FIGURE 9.6 A small section of the Heezen–Tharp 1957 physiographic diagram of the Atlantic Ocean. It defined the large-scale physiological provinces of the seafloor and highlighted its major physical features. Production of this map was heavily dependent on Cold War military funding and extensive support from Bell Labs, who were then laboring to install the first transatlantic telephone lines.
Source: US Library of Congress, Washington, DC. Simon Mitton

publication in any format of precise bathymetric data deeper than 300 fathoms (550 metres). To overcome this restriction, Heezen and Tharp developed techniques of physiographic mapping. Such an approach, illustrated in Figure 9.6, beautfully illustrates the dominant landforms and provinces of the ocean from an oblique perspective. This method of visualization made it easier to extrapolate morphology in regions where the depth data were sparse.

The first Heezen–Tharp physiographic map of the North Atlantic, drawn in pen and ink, was ready in draft form by 1957. The presentation of it was strongly influenced by Heezen's consulting contract with Bell Labs for the route of their new transatlantic *telephone* cable. This consultancy gave him access to the older records of cable breaks from the international *telegraph* firms in France and Britain. Through his work with engineers on the practical problems of cable laying, Heezen became skillful at using visualization to highlight particular aspects of the topography of the seafloor. Tharp used contours and shading to "show the terrain as it would look from a low-flying plane." This unusual perspective resembles many close-packed, jagged mountain ranges.[12] With a popular audience in mind, Heezen and Tharp adopted the Mercator projection, so familiar in every geography schoolroom. Heezen also wanted the map to expound the expanding Earth theory in the best possible light.

Bell Labs provided the funds that enabled Heezen and Tharp to draft the maps, and so it was natural that the Lamont team published in the *Bell System Technical Journal*.[13] Two years later, the Geological Society of America reissued the map with a monograph in which Ewing, Heezen and Tharp gave their interpretations. This was a capstone to seven years' labor by a huge team of scientists and mariners, born of more than 50 expeditions. Their monograph identified three distinct provinces: the continental margins, the ocean basin floor with its vast abyssal plains and the continuous mid-ocean ridge with its central rift. Critics of the map and its interpretation were soon on the case. A British oceanographer, Anthony S. Laughton, wrote to Heezen questioning the existence of the central rift.[9] Others dismissed the map as speculative and of marginal value owing to its limited number of profiles and soundings. Henry William Menard (1920–1986) at Scripps argued that cartography should be limited to carefully surveyed regions and that unhelpful speculation should be avoided.

Criticism of the map did not deter Harry Hess, who looked beyond the limitations of the interpolations and extrapolations made by the cartographers: of necessity they had had to engage in guesswork or intuition where the data were inadequate. The curious terrain of the ocean basins, dominated by an enormous mountain range with its prominent central valley bisecting the vast plains, greatly intrigued him. Hess also pondered deeply two baffling geological puzzles. One was posed by the marine fossils in the ocean floor sediments. The embedded fossils indicated that the oldest *oceanic* crust is younger than 180 million years old, whereas the oldest *continental* rocks date from 4 billion years ago. The second conundrum was that in the Pacific Ocean the accumulated sediments lying on the basement were only 1 kilometre thick, whereas the rate of sediment deposition would have amounted to 20 kilometres if sustained over the age of Earth. The evidence pointed to the Pacific basin being far younger than billions of years.

☉ THE GEOPOETRY OF HARRY HESS, 1962

In June 1960, Bruce Heezen visited Iceland, where the northern Mid-Atlantic Ridge rises above sea level. At that time, no one had directly observed the undersea range of mountains. When Heezen photographed the Central Island Graben (the floor of the rift valley) from the air, he was struck by its many tension cracks. Given the youthful age of the rocks at the base of the rift, he argued that magma was welling up from below the crust. He told Hess that he had observed Earth "coming apart at the seams," and he decided he was witnessing firsthand the consequences of Earth's expansion. In a popular account for *Scientific American*, he posited that Earth is

> neither shrinking nor remaining at the same size; rather it is
> expanding. If the earth were expanding and the continents
> remained the same size, additional crust would have to be formed
> in the oceans. This is apparently just what is happening in the mid-
> ocean rift valleys.

Heezen went on to explain that:

> Maurice Ewing and others hold that the ridge has its origins in
> convective currents within the earth. Heat-flow measurements
> show ... that heat is flowing from the interior of the earth much
> faster through the ridge and rift than through the ocean basins and
> the continents ... The upward and lateral movement of the
> currents would pull the rift apart and force material out through
> the crack.

He admitted that this proposed mechanism required the dismissal of paleomagnetic data that indicated the relative positions of the continents had changed. We'll come to that aspect in the next chapter.[14]

Heezen's vivid prose delighted Harry Hess, who was ever eager to explore the implications of new discoveries for the ocean crust. Hess adopted a neglected proposition of Arthur Holmes, who, as early as 1919, and in the final chapter of his major textbook *Principles of Physical Geology*, had suggested a mechanism for moving the continents. Warning that his ideas were "purely speculative," Holmes had imagined that rocks in Earth's interior, heated and melted by radioactivity, could rise buoyantly from the mantle toward the surface and then sink back again as they cooled. This was a convection model similar to the way in which warm air circulates in a room. Holmes' formed a mental picture that this convection could function as a conveyor belt, propelling and scrunching the crust sideways. More than any other geophysicist of his time, Holmes grasped the importance of convection as a possible physical explanation of heat loss and the cooling of Earth's interior. But as we have seen several times already, his fixist contemporaries could not embrace this mobile concept in the absence of hard evidence in its favor.

Hess was of a different generation. He was open to new ideas. By the mid-1950s, Hess had finally warmed to Holmes' model of convection cells in the mantle as a way of explaining the crustal buckling beneath island arcs – long arc-shaped chains of volcanic islands bounded on the convex side by a deep oceanic trench. In 1954, he boldly suggested that the Mid-Atlantic Ridge was a topographically elevated region lifted "by an upward convection current." His overall picture was one of the ridges being thrust up by heating in the mantle and oceanic crust being dragged down into the trench system. Implicit in this scheme was that ridges would evolve and that they are ephemeral over geological time. In terms of belief, he resisted mobilism right up until 1959, when he changed his mind because new results from rock magnetism were pointing to strong evidence that the continents had moved.[15]

Once Hess had bought into mobilism, he had to reexamine various features of the ocean basins. This he did in a preliminary manuscript rushed out in 1959, which quickly attracted attention. On July 6, 1959, he wrote to Vening Meinesz, saying, "It is becoming almost compelling to accept continental drift, something I never favoured." In December 1960, he followed up with a preprint entitled "Evolution of Ocean Basins," in which he eliminated the concept of the entire ridge being elevated by heat from the mantle. New data on the high heat flow from the crests of rising oceanic ridges encouraged him to develop a dramatic new synthesis that he promoted through the 1960 preprint. This preliminary account became widely cited and was finally published in 1962 as "History of Ocean Basins."[16] Like every scientist, Hess penned peer-reviewed prose to promote his progress on understanding the ocean basins. However, this paper was in a different register: he described it not as a paper but as an essay in geopoetry, in which he did not stray too far into the sublime realms of fantasy, but nevertheless wanted his readers to treat it as a romance, in which volumes of previously unrelated facts on the evolution of ocean basins would fall into a regular pattern that might suggest a close approach to a satisfactory theory.

His own words set the scene:

> The birth of the oceans is a matter of conjecture, the subsequent history is obscure, and the present structure is just beginning to be understood. Fascinating speculation on these subjects has been plentiful, but not much of it predating the last decade holds water.

There's a neat pun here (holding water), and his opening echoes Mark Twain's famous witticism from *Life on the Mississippi*:

> There is something fascinating about science. One gets such wholesale returns of conjecture out of such a trifling investment of fact.

Hess made the simplifying assumption that soon after the formation of Earth, heating by radioactivity caused a convective overturn of the interior, leading to the formation of an iron core, the mantle and the primordial continents. He pictured the excess heat flow of mid-ocean ridges as a consequence of mantle convection cells, with the high elevation of the ridges being due to thermal expansion from convecting cells. Hess envisaged that oceans opened out from their centers, with molten material (basalt) oozing up from Earth's mantle along the mid-ocean ridges to form a newly created seafloor spreading away from the ridge in both directions. As spreading continued, the older ocean floor was pushed further from the ridge, where it cooled and subsided to the level of the abyssal plain, which is approximately 4 kilometres deep. In his conclusions, we can read the following gems:

> The whole ocean is virtually swept clean (replaced by new mantle material) every 300 to 400 million years.

> This accounts for the relatively thin veneer of sediments on the ocean floor . . .

> and the relative absence of rocks older than the Cretaceous in the oceans . . .

> The continents are carried passively on the mantle with convection and do not plow though oceanic crust . . . Their leading edges are strongly deformed when they impinge on the downward moving limbs of the convecting mantle.

> The oceanic crust, buckling downwards into the descending limb [of the convection cell] is heated and loses its water to the oceans . . .

> The cover of oceanic sediments and the volcanic seamounts also ride down into the jaw crusher of the descending limb, are metamorphosed, and eventually are probably welded onto continents . . .

> Ocean basins are impermanent features, and the continents are permanent, although they may be torn apart or welded together and their margins deformed . . .

> The Earth is a dynamic body with its surface constantly changing . . .

Harry Hess, cloaked in the mantle of a poet (and why not?), was cautious about promoting his revolutionary scheme for the evolution of ocean basins. He knew that not all of his many assumptions would be found to be correct, but he sensed that he had cast an assortment of observations into a new order. He ended his geopoetry essay with the following comment: "Nevertheless it appears to be a useful frame-work for testing various and sundry groups of hypotheses relating to the oceans." There's a hint of the humility of Holmes here in the use of "hypotheses." On page 506 of *Principles of Physical Geology*, the caption to Holmes' figure 262 of ascending currents in the mantle begins: "Diagrams to illustrate a purely hypothetical mechanism for engineering continental drift."

◉ ROBERT DIETZ: THE SEAFLOOR IS
A CONVEYOR BELT

Before concluding this chapter, I shall acknowledge that Hess was not the only person who contributed to the seafloor spreading hypothesis. Robert S. Dietz (1914–1995) completed his graduate studies at Scripps, San Diego, in 1941, where he mapped seafloor canyons off the coast of California. After war service, he worked as a government scientist run-ning a seafloor research group at the US Naval Electronics Laboratory (NEL), San Diego. From 1946 to 1963, he made 10 cruises to the Gulf of California, mapping and scuba diving at the heads of submarine canyons. Diving gave him a unique point of view as an exploration geologist and led Dietz to persuade the US Navy to purchase the bathyscaphe *Trieste* for manned submarine research by NEL.

Dietz first chatted about seafloor spreading in a lunchtime conversation with a doctoral student at Scripps, Robert L. Fisher, in 1953. At that time, Dietz held a Fulbright Fellowship in Japan, where he was analyzing soundings of the North Pacific. He traced a sea-mount chain running northwest from Midway to the Kamchatka Trench, and he surmised to his younger colleague that these old submerged volcanoes "must be on some sort of conveyor belt." In 1954, Dietz proposed official designations for nine of the conveyor-belt guyots, naming them after Emperors of Japan. During the 1950s and 1960s, Dietz was an accomplished advocate of the dynamic properties of the ocean floor. The term "seafloor spreading" first appears in his *Nature* article published on June 3, 1961, with the title "Continent and Ocean Basin Evolution by Spreading of the Sea Floor": "The concept proposed here, which can be termed the 'spreading seafloor theory', is largely intuitive, having been derived through an attempt to interpret seafloor bathymetry."

At more or less the same time, Dietz and Hess independently articulated similar concepts of the seafloor spreading from the mid-ocean ridges. Hess is given priority for these ideas because in 1960 he widely distributed the preprint of what became published in 1962 as *History of Ocean Basins*. Furthermore, unlike Dietz, he had spent many years studying the mid-ocean ridges. The two models were not identical, however. Importantly, Dietz correctly suggested that convection currents cause the brittle lithosphere to ride on top of the plastic asthenosphere. On the other hand, Hess suggested that the base of the crust is driven by the currents. In his *Nature* paper, Dietz describes the ocean floor as the "exposed and outcropping limbs of convection." Dietz also made the correct call on the nature of the fracture zones that run at right angles to the ridges: he thought that they were the result of uneven convection in the mantle moving parts of the ocean at different rates. Hess felt that fracture zones were not related to the mid-ocean ridges. We know these fractures as transform faults.

Seafloor spreading can be described as the formation of new oceanic crust at spreading centers along the crest of mid-ocean ridges, where magma rises through fractures as the crust rifts and then cools to form new seafloor. Overnight, this theory made sense of a vast array of geological data from myriad sources. However, despite these successes, Hess and Dietz and others knew that, like Wegener and Holmes before them, incontestable geophysical evidence was essential to support seafloor spreading as the explanation of continental drift. Within a year or so, this would come from an unexpected line of inquiry: the surveying and mapping of the magnetic properties of the ocean floor. We shall return to this topic in Chapter 11.

REFERENCES

1. Matkin, J. *At Sea with the Scientifics: The Challenger Letters of Joseph Matkin*, P. F. Rehbock (ed.) (University of Hawaii Press, 1992).
2. Buckley, A. *Winners in Life's Race* (Edward Stanford, 1883).
3. Deacon, M. *Scientists and the Sea 1650–1900* (Academic Press, 1971).
4. Friedman, H. The legacy of the IGY. *Eos, Transactions American Geophysical Union* **64**, 497–499 (1983).
5. Dietrich, G. Einige morphologische Ergebnisse der "Meteor"-Fahrt Januar bis Mai 1938. *Annalen der Hydrographie und maritime Meteorologie* **67**, 20–23 (1939).
6. Hess, H. Geological interpretation of data collected on the cruise of USS *Barracuda* in the West Indies. *Eos, Transactions American Geophysical Union* **18**, 69–77 (1937).
7. Ewing, M. Gravity measurements on the USS *Barracuda*. *Eos, Transactions American Geophysical Union* **18**, 66–69 (1937).
8. Woodward, C. V. *The Battle for Leyte Gulf* (Macmillian, 1947).
9. Doel, R. E., Levin, T. J. and Marker, M. K. Extending modern cartography to the ocean depths: military patronage, Cold War priorities, and the Heezen–Tharp mapping project, 1952–1959. *Journal of Historical Geography* **32**, 605–626 (2006).
10. Tharp, M. Connect the dots: mapping the seafloor and discovering the mid-ocean ridge. In *Lamont–Doherty Earth Observatory of Columbia University, Twelve Perspectives of the First Fifty Years 1949–1999*, L. Lippsett (ed.) (Palisades, 1999), pp. 31–37.
11. Wertenbaker, W. *The Floor of the Sea* (Little Brown, 1974).
12. Bullard, E. The emergence of plate tectonics: a personal view. *Annual Review of Earth and Planetary Sciences* **3**, 1–31 (1975).
13. Elmendorf, C. H. and Heezen, B. C. Oceanographic information for engineering submarine cable systems. *Bell System Technical Journal* **36**, 1047–1093 (1957).
14. Heezen, B. C. The rift in the ocean floor. *Scientific American* **203**, 98–114 (1960).
15. Frankel, H. Hess papers, Princeton University. Letter to Vening Meinesz dated 6 July 1959. Harry Hess develops seafloor spreading. In *The Continental Drift Controversy* (Cambridge University Press, 2012), pp. 198–279.
16. Hess, H. History of ocean basins. In *Petrologic Studies: A Volume to Honor A. F. Buddington* (Geological Society of America, 1962), pp. 599–620.

10 Earth's Deep Dynamics Discovered

⊙ SEISMOLOGY DIGS DEEP TO THE CORE

The previous chapter described the genesis of a new interdisciplinary field, oceanography, which led to enormous advances in broadening our understanding of the history of the ocean basins. Seafloor spreading at the ocean central ridges and the ancient seafloor descending into the trenches at the ocean margins were the key conceptual breakthroughs for discovering the interior dynamics of the Earth system. But turning that imaginative *concept* of a dynamic Earth into a realistic package of *evidence*, as required by the scientific method, continued to be elusive in the 1950s and early 1960s. As is often the case in the history of Earth science, though, the convincing evidence eventually came from two unexpected but related lines of investigation: studies of rock magnetism and magnetic surveys of the seafloor. Investigation and follow-up of the phenomena in these fields turned out to be of critical importance for reconstructing the history of the motions of portions of the terrestrial surface. This is a tale of serendipity, in which research on the magnetic record fossilized in rocks finally led to the seal of approval for a brilliant concept: plate tectonics. I'll begin with the puzzle of how and where the geomagnetic field is generated. The properties of the geomagnetic field are observable from surface measurements of its direction, strength and variations. The major part of the surface field is generated by deep electrical currents in the core, although there are also small contributions from the magnetization of crustal rocks.

In Chapter 4, we noted that in 1600 William Gilbert's *De Magnete* had set out the postulate that Earth's interior contains a *permanent* magnet with two opposed magnetic poles situated at the geographic poles. Simplicity itself! In 1686, Edmond Halley added complexity when he conjectured that our planet's structure is an outer shell enclosing an inner *nucleus*, with a fluid medium in between. Halley's suggestion of a *nucleus* is the first mention in natural philosophy that Earth might have a core. Two centuries later, Emil Weichert (1861–1928) divided Earth's interior into two shells: a silicate mantle wrapped around a heavy *metallic* core.[1] He arrived at this opinion by noting that Earth's average density (5.6) is almost double that of the surface rock (about 3), implying that Earth cannot be composed entirely of rock.

Seismology, the scientific study of the seismic waves produced by earthquakes, contributed to our knowledge of the solid Earth from 1875, when Filippo Cecchi (1822–1887) built the first time-recording seismograph. John Milne (1850–1913) and his colleagues, professors of seismology in Japan, built the horizontal pendulum seismograph, which became the standard instrument for seismic observatories. Richard Oldham (1858–1936), who worked in the Himalayas for the Geological Survey of India, was the first to distinguish three distinct waveforms recorded on seismograms: primary (P), secondary (S) and tertiary (surface) waves. He grasped the possibility of discovering Earth's interior structure from seismic data on the arrival times of different waves from the same earthquake.[2] His analysis in 1900 of the travel times of seismic waves from the 1897 earthquake in Assam, India, prompted him to suggest that Earth has a dense central core that slowed the P-waves. Although his initial argument was unconvincing, he collected more data and refined his ideas for a landmark paper published in 1906 that conclusively established the existence of Earth's core.[3] He gave the depth to the core as 3900 kilometres. He did not, however, discover the core to be liquid, although in 1913 he wondered if the core might be in a "fluid or solid-fluid to the gaseous state," which kept the options open.[4]

Oldham's findings had an impact because they stimulated others to decode the signals in seismic data. Thus, in 1910, by studying changes in seismic velocities, Andrija Mohorovičić (1857–1936),

FIGURE 10.1 Memorial plaque in the Clementinum, Prague, Czech Republic, dedicated to Andrija Mohorovičić, who studied at the Charles University from 1875 to 1878. It marks the centenary of his discovery of the discontinuity that bears his name. He analyzed seismographs of the earthquake of October 8, 1909, the epicenter of which was 39 kilometres southeast of Zagreb, Croatia.
Source: Simon Mitton

subject of the bronze plaque in Figure 10.1, found the discontinuity between the crust and the mantle that now bears his name, abbreviated as the Moho. It is at a depth of about 10–12 kilometres under the ocean floor and 40–50 kilometres beneath the continents. In 1913, one of Weichert's students, Beno Gutenberg (1889–1960), discovered a major discontinuity in seismic wave velocities at a depth of 2900 kilometres. This marks the boundary between the mantle and the core, and Gutenberg made the first correct determination of the radius of the core. Thereafter, he continued to believe that the core was solid. That changed in 1926, when Harold Jeffreys reworked Gutenberg's analysis and found it could be interpreted to support a theory that the core is fluid. Inge Lehmann (1888–1993) further improved our understanding of the core in 1936, when she found the discontinuity that marks the boundary between the outer core and the inner core. Her inclination was that the inner core must be solid, but this hypothesis was not confirmed until the 1970s.

⊙ THE PHYSICISTS AND THE DYNAMO

In the first decades of the twentieth century, the merest hint of a high-pressure interior of liquid iron stirred the theoretical physicists. They were excited by discovery of a sharp structural discontinuity about halfway from the surface to the center: we know this as the core–mantle boundary. Geophysical ideas had coalesced around the concept of a *molten* iron core, wrapped up in a mantle of conventional rock. But what left the theorists baffled was how geomagnetism could fit into this worldview. The concept of a primordial permanent bar magnet had long been ruled out by Earth's great age, opening the way for competing theories about the source of geomagnetism. Albert Einstein, whose grasp of field theory and electromagnetism was immense, promoted a school of thought based on fundamental physics. In 1905, shortly after writing his famous paper on the special theory of relativity, he ranked the origin of the geomagnetic field as one of the most important unsolved problems in theoretical physics. As a fundamentalist, he pointed out that *if* the charges of the electron and the proton should be not absolutely identical in magnitude, the resulting asymmetry of electric charge distribution would make Earth a *permanent* magnet due to its axial rotation. By contrast, in 1919, the Irish mathematical physicist Sir Joseph Larmor (1857–1942) suggested that fluid motions in the core generated the geomagnetic field through a dynamo mechanism.[5] The solution to this intellectual puzzle of the geomagnetic field then became a contest between theoretical (Einstein) and applied physics (Larmor). When Harold Jeffreys finally established the fluidity of the core to everyone's satisfaction in 1926, he wrote succinctly, "There seems to be no reason to deny that the earth's metallic core is truly fluid."[6] For Jeffreys, this was an astonishing reversal of belief, made possible by his own calculations: as late as April 1925, he had given a lecture on the cooling of Earth in which he still continued to assume solidity for the interior.[7]

Walter M. Elsasser (1904–1991) pioneered the elaboration of the mathematical theory of a self-sustaining dynamo driving the geomagnetic field (Figure 10.2). Born in Mannheim, Germany, during his early career he studied physics successively in Göttingen, Leiden, Zurich, Kharkov, and Frankfurt. Because of his Jewish ethnicity, he abandoned Nazi Germany and sought refuge in the USA, where, in 1937, he obtained a research position in the Meteorology Department at the

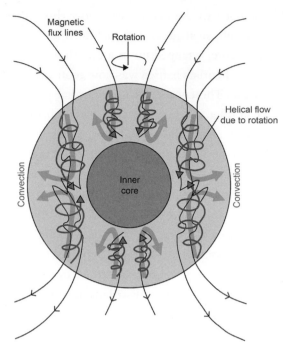

FIGURE 10.2 The principal features of a self-sustaining dynamo
mechanism for the geomagnetic field. Convection currents in the fluid
outer core are the result of rotation.
Source: Penny Wieser

California Institute of Technology (Caltech). His expertise in atmospheric physics led to defense work for the US Signal Corps
Laboratories, Fort Monmouth, New Jersey, where he initially investigated the propagation of electromagnetic waves. That research
sparked his deep interest in electrically conducting fluids and geomagnetism. By the late 1940s, he had figured out how an electrically
conducting core of liquid iron could sustain a long-lived magnetic
field. This theory is extremely technical, but the essential point is
that Elsasser seamlessly stitched together two distinct branches of
mathematical physics: hydrodynamics and electromagnetism.
Initially, his model had its skeptics, possibly because his extensive
use of vector calculus was way over the heads of most geologists.

Nevertheless, Elsasser laid the foundation for the theory of the turbulent convective geodynamo, which remains a highly sophisticated area of geophysical theory today. The first numerical models appeared in 1995: these accounted for a mostly dipolar field aligned with the rotation axis and for the reversals of polarity, which we shall now consider.

⊙ PALEOMAGNETISM AND THE HISTORY OF ROCKS

Earth's magnetic poles have reversed many times in geologic history. However, general acceptance of the evidence of reversals came only in the early 1960s, half a century after the distinguished astronomer George Ellery Hale (1868–1938) had discovered the 22-year cycle of reversals for the solar magnetic field.

In the mid-nineteenth century, a handful of continental geologists had noted that some rocks exhibit permanent magnetization acquired from the geomagnetic field. Laboratory experiments revealed that lava fired to red heat and then allowed to cool became magnetized parallel to the local field.[8] These results intrigued Bernard Brunhes (1867–1910), who was director of the Puy de Dôme meteorological observatory (Figure 10.3), perched at 1465 metres atop a large extinct volcano in the Auvergne Volcanoes Regional Park in central France. In 1905, he carefully investigated *in situ* an ancient lava flow that was magnetized in a direction opposite to that of the present-day magnetic field. He went on to describe the magnetization of several formations in the volcanic Massif Central, France. Unlike all previously documented rocks, the direction of magnetization recorded by both a Miocene basaltic lava flow and underlying baked clay was reversed compared to the present, indicating the magnetic North Pole had been close to the geographic South Pole at the time the lava was emplaced.[9] This was the first indication of the exciting possibility of recovering the long-term history of the geomagnetic field by revealing the fossil field frozen into lava at the time of its formation. Brunhes' strange result attracted little attention at first, possibly because there was no physical mechanism to account for the

FIGURE 10.3 Postcard (1910s) of three smartly attired sightseers descending the bumpy basalt flanks of the Puy de Dôme, Auvergne. The meteorological observatory, then directed by Bernard Brunhes, is shown in the top center.
Source: Simon Mitton

geomagnetic field. Fortunately, two geologists did take Brunhes seriously: Swiss glaciologist Paul-Louis Mercanton (1876–1963) and Japanese geophysicist Motonori Matuyama (1884–1958).

Mercanton was an Arctic explorer who intuited that if the reversals in rock were caused by the flipping of the geomagnetic field, then rock samples with reversed magnetization would surely be found globally. He set out to check this conclusion. Between 1910 and 1932, he curated lava specimens from Greenland, Spitsbergen, Iceland, the Faroe Islands and Australia, finding normal and reversed magnetization in both hemispheres. Matuyama did not take up the quest until April 1926, when he set out on a disciplined study of the hypothesis that some rocks had switched polarity. He chiseled out 139 Quaternary basalt specimens in Manchuria, Korea and Japan. His examination of their remnant magnetization led to a series of papers, the most important of which was published in 1929.[10]

Finding a clear correlation between polarity and stratigraphic position, he announced:

> According to Mercanton the earth's magnetic field was probably in a greatly different or nearly opposite state in the Permo-carboniferous and Tertiary ages as compared to the present. From my results it seems as if the earth's magnetic field in the present area has changed even to the opposite direction in comparatively shorter duration in Miocene and also Quaternary periods.

Unfortunately for the history of geophysics, this amazing discovery was not recognized until the mid-1950s, by which time a whole generation has passed.

⊙ PIONEERING GEOMAGNETISM AT THE CARNEGIE INSTITUTION

A survey of work on geomagnetism in the USA during the first third of the twentieth century takes us to the Department of Terrestrial Magnetism (DTM) of the Carnegie Institution of Washington, DC. Louis Agricola Bauer (1865–1932) was the first director from 1904 to 1929. He had studied civil engineering at the University of Cincinnati, then joined the US Coast and Geodetic Survey (USCGS), where he learned the practical aspects of geomagnetism. Bauer earned his doctorate at the University of Berlin in 1895, where he took courses in theoretical physics given by the luminaries of the Berlin School. The theoretical physicist, and later Nobel Laureate, Max Planck had been one of his thesis examiners. When Bauer returned to the USA, he had settled on his life's goal, which was to raise the profile of geomagnetism to the status enjoyed by other fields of physics. As head of DTM, Bauer conducted a full geomagnetic survey of the country and reorganized the network of magnetic observatories, as well as introducing a program of geomagnetic measurements on ships.

Bauer was motivated to deploy his wealth of experience on an international scale. That's because he concluded that no theory of geomagnetism could be developed until the magnetic elements (such as vectors, intensity and variability) of a significant portion of Earth's surface and oceans had been mapped over a comparatively short period of time. He vigorously began mapping the geomagnetic field of the entire Earth. During his tenure, DTM sponsored 130 magnetic expeditions to remote parts of the globe: at one stage, there were about 200 observers on land and at sea gathering data. DTM had two ships for ocean magnetic surveys. DTM chartered the brigantine *Galilee* from 1905 to 1908. She made valuable contributions to the DTM mission, but had shortcomings due to the magnetic material in its structure. Then, from 1909, the nonmagnetic, wooden-hulled yacht *Carnegie*, a seagoing observatory commissioned by Bauer, sailed around the globe many times taking magnetic measurements. She made a hazardous circumnavigation of Antarctica between 1918 and 1919, dodging numerous icebergs thanks to constant watchfulness. Her seventh cruise, following an extensive refit, was planned to last three years, but it was tragically ended prematurely while bunkering at Apia, Samoa. A gasoline explosion on November 30, 1929, severely injured the captain, who died within a few days. The cabin boy (an adult employee serving the captain) was lost overboard and several crewmen were badly injured. Within hours, the ship burned to the waterline. Carnegie officials never replaced her, and instead DTM changed course in the 1930s to less expensive research.

By the time of the tragic accident, Carnegie had gathered a huge amount of geomagnetic data for magnetic mapping, but the theorists were no closer to a detailed understanding of the *variable* geomagnetic surface field. From 1937, DTM supported investigation of the magnetic properties of sedimentary rocks. The following year, DTM brought online a sensitive magnetometer to measure all of the elements of Earth's magnetic field. Sediment cores taken in the north Atlantic Ocean and results from studying Pleistocene varved clays soon indicated a slow drift in Earth's magnetic field, with orientations that differed from the present field by up to 42°. The outbreak of war brought this research project to a standstill, and when it resumed in 1946, efforts became focused on using rock magnetism to differentiate conflicting ideas about the origin of the field. Meanwhile, American geologists continued to be rigidly opposed to continental movement. They distrusted paleomagnetic results as being intrinsically ambiguous because of magnetic reversals and uncertainties about the long-term stability of rock magnetism.

John Warren Graham (1918–1971) joined DTM in 1947 on a research fellowship while continuing as a part-time doctoral student at Johns Hopkins University. He had a strong background in geology and applied his ingenuity at instrumental development to build a spinner magnetometer that became DTM's workhorse for rock magnetism. With this high-sensitivity probe, he demonstrated that sedimentary rocks retain a direction of magnetization through long periods of geological time, even if the rock becomes folded. He suspected that paleomagnetism might offer clues as to whether or not the continents had drifted. His 1949 paper on the stability and significance of magnetism in sedimentary rocks is a classic work in the field of rock magnetism. In it, he suggested several lines for future research, including the following question: "Will the study of rock magnetism be useful in throwing light on questions regarding the large-scale movements of the crust, such as continental drift and polar wandering?"[11] This paper established paleomagnetism as a reliable field of inquiry concerning geomagnetism and Earth history. Graham continued in the field, but his DTM director, Merle Tuve (1901–1982), had become less supportive of research that touched on continental drift, on the grounds that it was geology not physics. In 1955, Graham published an impressive paper on the magnetizations of 343 samples of sediments from the USA. His analysis gave similar results to those already published by the British, apart from a significant difference in the orientation of the fossil field. He interpreted this as being due to slippage of the lithosphere relative to the axis of Earth's rotation. But by this stage he seems to have a suffered a loss of nerve: in 1957, his position at Carnegie was discontinued, and he moved to the Woods Hole Oceanographic Institute in Massachusetts.[12]

☉ APPARENT WANDERING OF THE
GEOMAGNETIC FIELD

Across the pond, paleomagnetism became established thanks to British physicist Patrick Blackett (1887–1974), who developed a sensitive magnetometer to test his novel ideas about the physics of magnetism.[13] In 1950 and 1951, he ran a long series of electromagnetic experiments using highly sensitive magnetometers of his own design. These experiments resulted in a valuable "instrumental study on the theory and use of the magnetometer." He had started a rock magnetism group in 1952, and its youthful members quickly found the new astatic magnetometer ideal for examining the direction of weak residual magnetism of sedimentary rocks. Two research teams were at soon work: Blackett's in Manchester, then subsequently (1953) at Imperial College London; and a second group directed by Blackett's former doctoral student Keith Runcorn (1922–1995), initially at Cambridge (1950) and then at Newcastle (from January 1956). Both groups made rapid progress in further improvements to magnetometers and to their application to the reconstruction of the history of the geographical position of the geomagnetic pole.

At Cambridge, Runcorn directed Jan Hospers (1925–2006), a Dutch graduate student with experience of stratigraphic fieldwork, to collect Icelandic lava samples for paleomagnetic investigation. At the end of the summer of 1951, Hospers returned to Cambridge with more than 600 samples. He began measuring them straightaway, using a magnetometer of his own design that was optimized for investigating strongly magnetized basalts. Hospers wrote seven papers on his findings. The first of these is seminal because its data offered extensive support for the concept of polarity reversals of the geomagnetic field.[14]

At the spring 1953 meeting of the American Geophysical Union, Runcorn gave his keynote speech on paleomagnetism. Although many in the audience were unfamiliar with paleomagnetism, Runcorn wowed them with his dramatic and articulate presentation of his graduate student's research on reversals in young basalts. For good measure, Runcorn added results by two more of his doctoral students, Edward "Ted" Irving (1927–2014) and Kenneth M. Creer (1925–2014). Irving carried the fieldwork and analysis on the magnetization of sedimentary sandstones, the older one dated at ~700 Ma (Torridonian) and the less ancient one dated at ~360–415 Ma (Old Red Sandstone). Both investigations covered a far longer time frame than that of the young basalts studied in the first half of the twentieth century. Runcorn related how the magnetizations of these ancient sediments were unusually stable, although he cautioned that experience was essential for being able to select suitable samples. He and the pair of students had found directions of magnetization that were systematically tilted by as much as 90° when compared to the present geocentric axial pole. These sensational findings implied that either the internal dynamo was bucking and twisting through a right angle, thus shifting the magnetic pole by large distances, or the British Isles had moved thousands of kilometres relative to the present axis of rotation. These were startling discoveries, with startling implications, one way or another.

Runcorn then linked three hypotheses in order to get a grip on these bizarre results. The first hypothesis was that, when averaged over thousands of years, the geomagnetic pole at Earth's surface is a geocentric dipole along its axis of rotation. To this he added a second hypothesis: that the geomagnetic pole has reversed its polarity many times in the past. And his third hypothesis was that polar wandering had occurred because Earth has toppled over gradually throughout geological time. In conclusion, he added a critically important rider: that since the rocks had been collected from Britain only, it was not possible to exclude continental drift as the cause of the change in the pole position.[15]

For Carnegie's John Graham and the paleomagnetic team from Princeton University who were in the auditorium, Runcorn's assertion that Earth's magnetic field had been flip-flopping throughout geological history was patently absurd: their hypothesis was that the ferromagnetic minerals in rocks had self-reversed by an unknown mechanism since the time of their emplacement. Runcorn's riposte was a call for more data from several continents. Hospers was particularly dismissive of Graham's work because of the mounting evidence that permanent magnetization of sedimentary rocks was unstable according to Irving.[16] The debate wore on for a further 15 years. At the collegial University of Cambridge, support for drift was absent. Irving's PhD thesis was failed in 1954, and the future Fellow of the Royal Society (1979) had to console himself with the lesser degree of an MSc. With no postdoc on offer at Cambridge, Irving moved in November 1954 as far away as possible: to the Australian National University, Canberra. There he began work on constructing a polar wandering path for the continent of Australia. In January 1956, Runcorn, too, deserted Cambridge, for the chair in physics at Newcastle University. A decade later, when Irving was awarded the higher doctorate of ScD by Cambridge on the basis of his peer-reviewed papers, he said that his failed PhD thesis had been "too skimpy." He was being extremely polite, of course.

⊙ DRIFTING, NOT WANDERING

The UK groups adopted a global approach to data collection by requesting specimens from North and South America, Africa, India and Europe. The new discipline advanced as a well-populated field of keen participants contributed to rapid progress from the mid-1950s. During this period, continental drift steadily moved ahead of polar wandering as the preferred explanation of the paleomagnetic data. Runcorn suddenly switched his support to drift shortly after leaving Cambridge. It seems that Irving and Creer had convinced him to do so after they had compared paleomagnetic results from Britain with their data recently acquired in Arizona and Utah.

In 1956, Runcorn published two continental drift and polar wandering papers in minor journals, *Proceedings of the Geological Association of Canada*[17] and *Geologie en Mijnbouw*. In these short papers, he finally accepted drift between Europe and North America.[18] Irving sent a paper on the comparison of polar wander paths, in which he touched on "relative movement of the continents," to the *Journal of the Geological Society of Australia*. Stung by a sharp rejection of it, he next submitted a revision to *Pure and Applied Geophysics*, which accepted it right away.[19] Half a century later, this little contribution was finally recognized for its great importance and was republished as a classic paper.[20] What the trio of papers illustrated was that the polar wander paths for Europe and North America had similar shapes, but the paths parted company as one went further back in time. It was shown that each of the five continents had a different path for (apparent) polar wandering, and that those differences were consistent with Wegener's hypothesis.[21] Once Runcorn switched sides, he became an effective and passionate advocate for continental drift.

Historically, the philosophical puzzle is why the continental drift solution proposed by the trio of talented geophysicists did not catch on right away.[22] They had enough evidence in the bag for a landmark paper in *Nature* or *Science*. Instead, they published in second-tier journals and avoided provoking their opponents with "continental drift," choosing instead neutral expressions such as "relative movement of the continents." Was the fault too much British reserve, perhaps? It was an object lesson in how not to publish research. Indeed, Irving waited until 1959 to claim that polar wander paths should be interpreted as strong support for continental drift.

Toward the end of 1956, Blackett, along with young colleagues John Clegg (1913–1987) and Peter Stubbs, published an analysis of rock magnetic data that had been collected (mostly by others) from five continents in the 1950s. In their introduction, the three *physicists* noted a wide divergence of opinion among *geologists* on the question of past movements of the continents, adding that "on balance most geologists appear to have lost faith in *continental drift as a working hypothesis*" (emphasis added). Blackett's group had mined data sets already in the public domain, reducing them to a consistent format to reveal for any given site the angle of divergence between the direction of the ancient field and that of the local dipole field.[23] This divergence steadily increased with geological age, just as Runcorn's group had found. They concluded:

> Some of the most striking results are as follows: all four of the land masses studied [Europe, North America, India and Australia] have during the last 200 My ... been moving steadily northwards with velocities of between 0.2 and 0.8° of latitude per million years. In the last 300 My, the mean latitudes of western Europe and North America have [moved] from the equator to their present positions ... In azimuth Europe has rotated about 50° clockwise ... Although the data are still scanty, India has apparently moved farther and faster than any other continent.

Once again, the impact of truly astonishing results was softened by typically British understatement: "The drift of continents across the earth's mantle" is the most plausible way to explain the rock magnetic data, but is a "tentative assumption" hinging on the reliability of paleomagnetism. By this stage, it was clear that each continent had its own path for "polar wandering," and that those differences were consistent with Wegener's hypothesis that the supercontinent Pangaea had started to break up 200 million years ago. Nevertheless, there continued to be an item of unfinished business in geology and geophysics: establishing an absolute chronology for the geomagnetic reversals. I take this matter up in the next chapter.

REFERENCES

1. Weichert, E. Über die Massenverteilung im Innern der Erde. *Nachrichten von der Gesellschaft der Wissenschaften zu Göttingen, Mathematisch-Physikalische Klasse* **1897**, 221–243 (1987).
2. Oldham, R. D. On the propagation of earthquake motion to great distances. *Proceedings of the Royal Society of London Series I* 66, 2–3 (1899).

3. Oldham, R. D. The constitution of the interior of the Earth, as revealed by earthquakes. *Quarterly Journal of the Geological Society* **62**, 456–475 (1906).

4. Brush, S. G. Discovery of the Earth's core. *American Journal of Physics* **48**, 705–724 (1980).

5. Larmor, J. How could a rotating body such as the Sun become a magnet? In *Report of the British Association for the Advancement of Science* (1919), pp. 159–160.

6. Jeffreys, H. The Earth's thermal history and some related problems. *Geological Magazine* **63**, 516–525 (1926).

7. Jeffreys, H. The cooling of the earth. *Nature* **115**, 876–877 (1925).

8. Brown, M. The early history of geomagnetic field reversals. *Institute for Rock Magnetism* **20**, 1–10 (2010).

9. Brunhes, B. Récherches sur la direction d'aimantation des roches volcaniques. *Journal of Theoretical and Applied Physics* **5**, 705–724 (1906).

10. Matuyama, M. On the direction of magnetisation of basalt in Japan, Tyosen and Manchuria. *Proceedings of the Imperial Academy* **5**, 203–205 (1929).

11. Graham, J. W. The stability and significance of magnetism in sedimentary rocks. *Journal of Geophysical Research* **54**, 131–167 (1949).

12. Doel, R. E. Memorial to John Warrant Graham 1918–1971. *Geological Society of America Memorials* **3**, 105–108 (1971).

13. Nye, M. J. Temptations of theory, strategies of evidence: P. M. S. Blackett and the earth's magnetism, 1947–52. *The British Journal for the History of Science* **32**, 69–92 (1999).

14. Hospers, J. Reversals of the main geomagnetic field, part I. *Proceedings of the Royal Netherlands Academy of Science, Series B* **56**, 467–476 (1953).

15. Creer, K. M., Irving, E. and Runcorn, S. K. The direction of the geomagnetic field in remote epochs in Great Britain. *Journal of Geomagnetism and Geoelectricity* **6**, 163–168 (1954).

16. Hospers, J. Reversals of the main geomagnetic field, part III. *Proceedings of the Royal Netherlands Academy of Science, Series B* **57**, 112–121 (1954).

17. Runcorn, S. K. Palaeomagnetic comparisons between Europe and North America. *Proceedings of the Geological Association of Canada* **6**, 163–168 (1956).

18. Runcorn, S. K. Palaeomagnetism, polar wandering, and continental drift. *Geologie en Mijnbouw* **18**, 253–256 (1956).

19. Irving, E. Palaeomagnetic and palaeoclimatalogical aspects of polar wandering. *Pure and Applied Geophysics* **33**, 23–41 (1956).

20. Frankel, H. R. Edward Irving's palaeomagnetic evidence for continental drift. *Episodes* **37**, 59–70 (2014).

21. Irving, E. The palaeomagnetic confirmation of continental drift. *Eos, Transactions American Geophysical Union* **69**, 994–1014 (1988).

22. Livermore, R. *The Tectonic Plates are Moving!* (Oxford University Press, 2018).

23. Blackett, P. M. S., Clegg, J. A. and Stubbs, P. H. S. An analysis of rock magnetic data. *Proceedings of the Royal Society of London. Series A. Mathematical and Physical Sciences* **256**, 291–322 (1960).

II Reversals of Fortune

⊙ DATING GEOMAGNETIC REVERSALS

The outstanding issue with reversals concerned the fidelity of the spreading ocean crust as a magnetic tape recorder: some geologists asked themselves if the reversals could have occurred spontaneously in the rocks sampled. This conundrum could be resolved by obtaining accurate dates of formation for groups of rocks with normal and reversed polarity, and so determining whether the rock magnetism reversed in sync or not. In the 1950s, uranium–lead (U–Pb) radiometric dating was not accurate enough for younger rocks because the half-life of uranium is so long, which meant the quantity of lead was insufficient for analysis by the technology then available. By the end of the 1950s, two groups of geochemists, one at the United States Geological Survey (USGS), Menlo Park, California, and the other at Australian National University, had developed the techniques of potassium–argon (K–Ar) dating to the stage where it could be used on young rocks, such as lava flows, with ages of hundreds of thousands to a few million years.

Allan V. Cox (1926–1987) became a graduate student in geophysics at Berkeley in September 1955, where he learned paleomagnetism alongside Richard Doell (1923–2008). Both were taken on by John Verhoogen (1912–1993), a Belgian–American geophysicist who was a pioneer in applying isotope geochronology to paleomagnetism. Verhoogen became highly influential in the development of the modern view of Earth's dynamic interior. He was the inspiration for a generation of scientists who applied physics and chemistry to unlock the inner secrets and contents of Earth. When Verhoogen learned about Irving's determination of a paleopole for

India, he suggested Cox should obtain a paleopole for North America, for comparison purposes.[1] Cox duly collected 57 samples from 8 individual lava flows of the Siletz River series in northwestern Oregon. His interpretation yielded a paleopole 30° northeast of the average pole found by Blackett's group, who had analyzed samples from the Deccan Traps in India.[2] Cox then concluded that accurate dates would be needed to interpret the past directions of remnant magnetism from different continents. He remained a fan of polar wandering. For his next excursion, he collected samples from Snake River Plain, Idaho, a monotonous landscape of lava flows 560 kilometres long and 70–110 kilometres wide. Here he unearthed normal and reversed rocks, which led him to an important discovery: the stratigraphic position of the normal–reversed horizon agreed with the position found by others working on different continents. That finished off for good the concept of rocks self-reversing: evidently, the solidifying lava did capture the local field at the time of formation. Henceforth, he decided to work with Doell on reversals of Earth's magnetic field. In 1960, Cox and Doell crafted a monumental review for the Geological Society of America (GSA) that provided a comprehensive summary of paleomagnetism up to the end of the 1950s. Although this well-written review by two little-known scientists failed to capture excitement within the flurry of recent results,[3] it was hugely influential because of the prestige and wide circulation of GSA publications. Cox and Doell still favored polar wandering, but without entirely rejecting continental drift. That is unsurprising: the two authors were early-career scientists who were schooled in the fixist tradition and aware that there was no safe middle ground in the controversy.[4]

The main benefit for science that arose from this flawed review was its impact on the community. Although Doell's supply of offprints was gone in no time, the leading mobilists in Britain, South Africa, Australia and the Soviet Union were highly critical. Nevertheless, the fixists' ball was still in play, but only just. Doell and Cox knew that they were onto something big, and they had several projects in mind, one being the determination of the time intervals between successive reversals of the geomagnetic field.

Late in 1962, they recruited Brent Dalrymple, fresh out of graduate school at Berkeley, where he had learned about isotope dating techniques from John Hamilton Reynolds (1923–2000), the pioneer who designed and built the first all-glass mass spectrometer for isotope analysis of the noble gases. Dalrymple's expertise with the Reynolds gas spectrometer led directly to the trio becoming highly acclaimed for their epochal work on the timing of recent reversals. Initially, they latched onto the idea that the reversals might be periodic (as is the case for the 11-year solar cycle), but on a much longer timescale. Their studies of the K–Ar ages and polarities of 64 volcanic rocks from North America, Hawaii, Europe and Africa were decisive in demonstrating that reversals of the geomagnetic field were the cause of reversals in rocks, and they had absolute dates for the reversals. Their paper illustrated four major reversals and two short-term events over a period of four million years (Figure 11.1). Our present period of normal polarity began 0.8 million years ago and is named the Brunhes period. Its predecessor of reversed polarity is the Matuyama anomaly (0.8–2.6 Ma). In their conclusion, the authors stated that magnetohydrodynamic theory had reached the point of development where it could account for the pattern of reversals.[5]

⊙ MAPPING THE MAGNETIC SEAFLOOR

I have already recounted how ocean cruises after World War II used sonar to map the topography of seafloor structures. By also towing a magnetometer a few hundred metres behind the ship at a depth of 15 metres or so, expeditions revealed the magnetic topography of the seafloor, and in particular the magnetic anomalies recorded by its rocks. The total field measured by the magnetometer is the sum of the global field and the local field produced by magnetic minerals in the surface rock. The strength of the global field ranges from 25,000 to 65,000 nanoteslas (nT), with the local field contributing in the range ±500 nT.

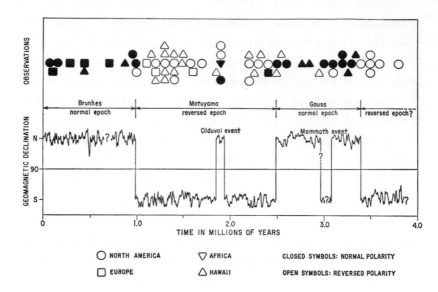

FIGURE 11.1 Cox, Doell and Dalrymple plotted the magnetic polarities and potassium–argon ages of 64 volcanic rocks. Their paper gave absolute ages for four major events, and that was convincing evidence that paleomagnetic reversals in rock reflected reversals of the geomagnetic field. *Source: Science*, **144**, 1541 (1964). American Association for the Advancement of Science

Ronald George Mason (1916–2009), a lecturer in geophysics at Imperial College London, was a pioneer who used towed magnetometers that he had re-engineered to great effect. The map of the magnetic survey off the West Coast of North America that he and his laboratory technician Arthur Datus Raff (1917–1999) published in 1961 would provide the long-sought legitimization of the theory of continental drift and plate tectonics, although it did not seem like that at the time. The story behind their publication of the map is another saga of risk-averse researchers being reluctant to release inexplicable discoveries.[6] In 1952, when Mason took sabbatical leave from Imperial College to spend a year at Caltech, he casually asked Russell Watson Raitt (1907–1995) at the Scripps Institution of Oceanography, San Diego, if anyone had thought of investigating the magnetic anomalies associated with oceanic structures by towing a magnetometer behind a survey ship. Roger Revelle (1909–1991), then director of

Scripps, overheard this question and quizzed Mason on whether he would like to be the "Scripps magnetometer man." Mason jumped at this opportunity. By 1955, Scripps had developed a suitable instrument by converting an airborne magnetometer for use at sea by towing it in a nonmagnetic, fishlike container. Getting the first marine magnetometer on board the US Coast and Geodetic Survey (USCGS) survey vessel *Pioneer* proved to be difficult. Its sponsors, the US Navy Hydrographic Office, were concerned that towing a magnetometer would literally be a drag on their hydrographic surveys. None of their advisors could see the point of a magnetic survey, and the USGS had pronounced it a waste of money. In the event, one of the most significant geophysical surveys ever made was wholly paid for by a small amount of discretionary funding available to Revelle.[7] With that support, Raff and Mason undertook a two-year series of monthly cruises on *Pioneer*. To begin with, their magnetometer data were so erratic that they could not sketch out any contours of magnetic anomalies. Some 50 years later, Ron Mason remembered that:

> The initial results were discouraging ... we might easily have abandoned the whole operation. But as the data accumulated, the nature of the field began to emerge: it was dominated by bands of north–south trending contours that extended the full [160 kilometres]. ... This was a period of great excitement because nothing like it had ever been observed before on land or at sea. There was no longer any question of abandoning the survey.

Indeed, it continued for a further 11 monthly cruises. When the survey was completed in October 1956, the tedious business of reducing the data from roll after roll of 7-metre pen-and-ink records was already underway. In 1958, Mason published an initial map of the magnetic anomaly survey from 32°N to 36°N (Figure 11.2). A distinctive north–south linear pattern is clearly evident, as is the Murray Fracture Zone (a transform fault).

In late 1956, Scripps diverted Mason to head its major International Geophysical Year (IGY) project in the equatorial

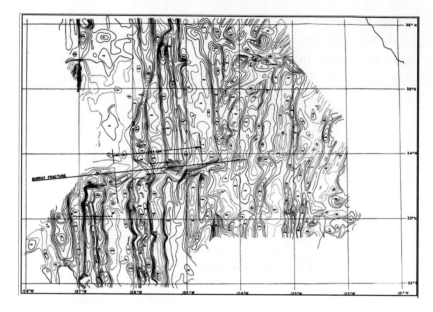

FIGURE 11.2 Publication of this map in the first volume of the fledgling *Geophysical Journal of the Royal Astronomical Society* was an obvious choice for Mason, who was lecturer in geophysics at Imperial College London, but the journal did not then command the immediate attention of geophysicists overseas.
Source: Geophysical Journal of the Royal Astronomical Society, **1**, 320–329 (1958)

Pacific, where temporary magnetic observatories were commissioned. Meanwhile, back in California, responsibility for continuing with the marine magnetic surveys aboard the Scripps research vessels *Horizon* and *Spencer F. Baird* passed to renowned geophysicist Victor Vacquier (1907–2009), who was hired in 1957. Vacquier was born in St. Petersburg, Russia, in 1907. In 1920, at the midpoint of the Russian Civil War, which had followed the two Revolutions of 1917 and resulted in the deaths of millions of citizens, Victor and his family escaped on a one-horse open sleigh across the frozen Gulf of Finland, en route to the safe haven of Helsinki. In 1923, he and his mother emigrated to the USA, where he received an electrical engineering degree at the University of Wisconsin in 1927. In 1940, he invented the fluxgate magnetometer, and in a career spanning seven decades received numerous high honors in geophysics. Vacquier extended the survey by running on three east–west lines more than 2000 kilometres in length. Mason learned a lot about magnetic surveys from Vacquier. Mason and Raff submitted papers on the magnetic anomalies to the GSA on June 3, 1960,[8] and December 16, 1960.[9] These were published back to back in the *Geological Society of America Bulletin* in August 1961: Figure 11.3 shows their renowned iconic map of magnetic stripes is on page 1268. Commenting on the stripes, Mason and Raff wrote:

> These north–south linear features have aroused considerable interest and speculation. ... There is as yet no satisfactory explanation of what sort of material bodies or physical configurations exist to give the very long magnetic anomalies.

In 2001, Mason outlined how he had tinkered with simple models and showed that "they could be explained by shallow, slablike structures, immediately underlying the positive stripes ... but there was no plausible geological model to support such structures." Mason was familiar with the hypotheses of seafloor spreading and magnetic reversals from his contacts back in London, but he was on the Pacific Coast where there remained considerable skepticism in 1960. Furthermore, no detailed magnetic survey spanning an ocean ridge was thought to be available. For two years, the "zebra stripes" remained unexplained in the open literature.

⊙ LAWRENCE MORLEY'S "EUREKA!" MOMENT

Raff and Mason's zebra map first appeared in August 1961. I literally freaked out when I saw it! I had been studying aeromagnetic maps from all over the world ... and had never seen such a regular linear pattern of positive and negative anomalies stretching for [1000 kilometres]. They might just as well have been features on Mars.

That was how Lawrence Whittaker Morley (1920–2013) described his electrified reaction to the map.[10] Morley was a Canadian geophysicist who became director of the Geological Survey of Canada in 1969.

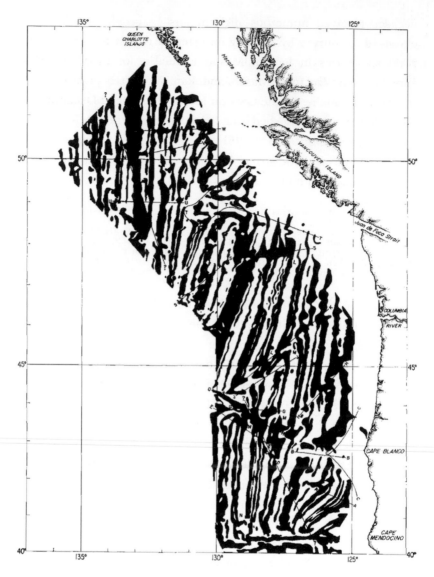

FIGURE 11.3 The celebrated index anomaly map of the total magnetic field produced by Raff and Mason. To uncover this spectacular pattern of anomalies, Mason reduced a vast amount of data acquired by a towed fluxgate magnetometer employed on many cruises. The total area surveyed was about 590,000 km^2.

Source: Geological Society of America Bulletin, 72, 1267–1270 (1961)

His introduction to rock magnetism was under John Tuzo Wilson (1908–1993) of the University of Toronto, whose own doctoral research at Princeton had been overseen by young Harry Hess. Morley was at the spring 1953 American Geophysical Union (AGU) meeting, where he was greatly surprised by Runcorn's flamboyant evocation of polar wandering and field reversals as the solution to the geomagnetic conundrum. When John Graham exclaimed that he disagreed with Runcorn on reversals, Wilson elbowed Morley, who leapt up and expounded his thesis results on the instability of magnetism in igneous rocks. For the next few years, Morley was busy with negative anomalies across the vast Canadian Shield. The Raff and Mason map set him thinking, a process that continued for a couple of years until he spotted Robert Dietz's 1961 *Nature* paper on seafloor spreading. That triggered a "eureka!" moment.

In a flash, Morley pictured mantle convection currents rising under ocean ridges, then traveling horizontally under the ocean floor, to sink at the ocean troughs. Such currents would lead to upwelling rock becoming magnetized in the direction of the geomagnetic field at the time as it spread out from the ridge and cooled below the Curie point. If Earth's field reversed in the meantime, while the process of spreading to make way for new upwelling rock continued, then a linear magnetic pattern of the type observed could be explained. It was an explanation that only required a convection cell with an axis of rotation at least as long as the anomalies, with a direction of travel stretching from ridge to trough, plus frequent reversals of Earth's magnetic field. Morley suggested that there had been as many as 180 reversals since the Lower Cretaceous and gave a convection current rate of 3.5 cm per year based on the width of a complete positive and negative cycle, as evidenced by the maps of Mason and Raff. Forty years later, Morley claimed: "I never had any doubts about the concept ... it locked together ... the theories of continental drift, seafloor spreading, and periodic [reversals]."[10]

Morley put these thoughts together in the form of a short letter to *Nature* in February 1963. Two months later, he received a fustian letter of rejection stating that because they did not have room to print it: "... the Editors present their compliments and beg to inform you that, etc. ..." In those days, *Nature* had an enormous backlog of papers and was at a low point in terms of its academic positioning as a science *magazine* rather than a serious *journal* of record.[11] Morley promptly submitted to the *Journal of Geophysical Research* (*JGR*) in April, but heard nothing until the end of August, when he received a boilerplate rejection letter along with feedback from a referee: "[Morley's] idea is an interesting one I suppose but it seems most appropriate over martinis rather than in *JGR*."[11] While Morley was wrestling with thoughts of publishing elsewhere, on September 7, 1963, *Nature* published the same idea in the celebrated article by Vine and Matthews.

⊙ FRED VINE'S "EUREKA!" MOMENT

Fred Vine freely admitted that the pivotal contribution he made to geophysics during the period from 1963 to 1966 was a classic case of being in the right place at the right time:

> I was lucky ... I manoeuvred myself into an area that struck me as being fertile ground for a possible breakthrough. There is little doubt, however, that the intellectual environment in the Department of Geodesy and Geophysics at Cambridge ... was an ideal spawning ground for such an idea.[12]

Vine remembers a fascinating diagram in a textbook of physical geography showing the approximate fit of the coastlines of South America and Africa. The caption stated that there was a hypothesis that these two landmasses had broken off from a supercontinent, but there was no way of testing the theory. This struck him as an issue that would need resolution before conducting a meaningful study of the history of Earth. He continued with his interest in physical geography during his final years in high school, where he won a state scholarship and admission to St. John's College, Cambridge, to read natural sciences. In his third and final year, he specialized in geology (with an emphasis on mineralogy and petrology) by attending lecture courses given by (Sir Edward) Teddy Bullard and Maurice Hill (1919–1966). Both were influential in his decision to undertake research in the subdiscipline of marine geology and geophysics. In January 1962, Harry Hess was in Cambridge as the distinguished guest speaker at a forum organized primarily by undergraduates on the theme of the evolution of the North Atlantic. In his lecture, he summarized the then-present state of marine geology and geophysics in terms of mantle convection and seafloor spreading. For Vine, it was "an inspiring and exciting synthesis." The following month, Vine was at the natural sciences club in Cambridge when Blackett spoke on paleomagnetism and invoked continental drift as being axiomatic in terms of accounting for polar wandering.

The next advance in Fred Vine's career came in April 1962, when he was awarded a doctoral studentship to work on marine magnetic data under the supervision of Drummond (Drum) Matthews (1931–1997), a research fellow of King's College (Figure 11.4). Drum was running magnetic survey lines between Africa and India on board HMS *Owen*, a naval hydrographic ship. Magnetic data were recorded nonstop (apart from the frequent breakdowns, occurring on average every four hours) using a homemade proton magnetometer built by Maurice Hill. Matthews returned from the Indian Ocean with a large amount of data on punched paper tapes.

FIGURE 11.4 Frederick Vine and Drummond Matthews, 1970.
Source: Department of Earth Sciences, University of Cambridge

The data included a detailed series of depth and magnetic profiles across the crest of the Carlsberg Ridge, so named because its discovery was made by the *Dana* expedition (1928–1930), a scientific mission funded by the Carlsberg Foundation. At the time, this survey of 6800 km² of seafloor was the largest and most detailed survey of a known mid-ocean ridge, with line spacings of 1.5–3.0 km. The process of reduction of the data to tease out the magnetic anomalies was not straightforward. Matthews preferred to play around with physical models based on simple shapes, such as slabs, where the geometry is defined by one or two parameters and subject to simple scaling laws. Vine realized this would not do for large-scale surveys with complex geometry. Instead of the analogue approach, he decided to try computer modeling. He already had knowledge of programming in two dimensions, but the Carlsberg Ridge data would require three dimensions. Accordingly, he sought the advice of Kanthia Kunaratham (1934–2015), who had just completed a doctorate under Ron Mason at Imperial College, in the course of which he had developed a program for interpreting the magnetization of the seamounts that Mason and Raff had surveyed in the 1950s. Back in Cambridge, Vine crafted a version of Kunaratham's program to run on Edsac 2 (one of the world's first mainframe computers) at the Cambridge Mathematical Laboratory. Then he set about building digital models.

To his delight, he found that the data concurred with two of his pet theories. The first was that rock on the ocean floor had spilled out in volcanic eruptions at the ridges and was then transported as if on a conveyor belt, as had first been suggested by Holmes in 1931, revived by Hess in the 1950s and promoted anew by Dietz in 1961, as noted in Chapter 9.[13] And the second was that the north and south magnetic poles switched periodically, reversing the field everywhere on Earth. By joining these two hypotheses together, Vine and Matthews picturesquely imagined new crust being created, with lava rich in magnetite capturing the direction of the local magnetic field as it cooled and solidified, and it then being transported not so much on a conveyor belt but more like a giant magnetic tape: "giant" meaning solid rock kilometres thick and a couple of hundred kilometres long. So far, so good, but if it only recorded small fluctuations in a uniform magnetic signal, not much could be learned. But the beauty of the scheme was that the field reversals would result in a sequence of north–south polarity intervals, creating something akin to a bar code from which one could read off Earth's history. Reversals were still not accepted by everyone in 1963, but Vine's interpretation of the survey was that 50 percent of the oceanic crust might be reverse magnetized. The Vine–Matthews (V–M) hypothesis elucidated the formation mechanism that led to the striking stripy scheme of magnetic anomalies.[14]

⊙ PLAYERS IN THE ENDGAME, 1963–1968

Nature handled the Vine and Matthews paper with alacrity, publishing it on September 7, 1963. Although Lawrence Morley never complained publicly about *Nature* and *JGR* rejecting his paper, there can be no doubt that he had been ill-served by his journal editors. Vine did not become aware of any of this until the late 1960s. In textbooks today, the concept is referred to as the Vine–Matthews–Morley (VMM) hypothesis. Here, I use the V–M format in its historical contexts.

The radical hypothesis did not gain traction straightaway – far from it. At the Royal Society symposium on continental drift held on March 19–20, 1964, Vacquier was the sole speaker who mentioned the V–M hypothesis, dismissively describing this "attractive mechanism [as] probably not adequate to account for all the facts of the observations." Malik Talwani, marine geophysicist at Lamont, deprecated it as a "startling explanation." That remark encapsulated the dictatorial attitude of most researchers at Lamont, which remained a citadel of constancy concerning the permanence of continents and oceans.[15] At St. John's, Cambridge, the avuncular Harold Jeffreys mildly mocked the musings of his college's doctoral student, Fred Vine. Maurice Hill, the departmental expert in Cambridge on marine magnetometry, was such a stickler for factual evidence that he brushed aside Vine's idea as sheer speculation. However, geopoet Harry Hess praised the V–M hypothesis as a "very *fruitful* idea introduced … by *Vine* and Matthews" (emphasis added).[16] Fred felt that 1964 was a fallow year for the V–M hypothesis, in part because he had to concentrate "on producing more substantial, or at least conventionally acceptable, material for my thesis." Perhaps he was mindful of Ted Irving's failed thesis of 1954. Matthews and Vine did team up with the petrologist Johnson R. Cann to coauthor two papers that discussed the geology of the Carlsberg Ridge in a way that gave more support for V–M hypothesis.[17,18]

The next year, Harry Hess and Tuzo Wilson spent several busy weeks in Cambridge working closely with Vine. Their meeting became the pivotal point of the entire mobilism debate. It was a most productive time for Wilson: he developed the idea of transform faults, and the trio found strong support for the V–M hypothesis.[19] A transform fault in geology or oceanography occurs when two plates slide past each other horizontally but without creating or destroying lithosphere. This type of fault begins at one plate boundary and ends at another plate boundary. Most transform faults are oceanic and are associated with the jagged parts – or fracture zones – of the ocean ridge system. Once Wilson had invented this idea, he applied it to the worldwide system of ridges and trenches. He realized that the San Andreas Fault, where southern California (Pacific plate) is creeping northward relative to the rest of California (North American plate), is a transform fault. Moving further north, off the coast of Washington State and Oregon, his logic indicated that there should be a short length of ridge in the Strait of Juan de Fuca connecting two transform faults. While Wilson was outlining his thinking to Hess and Vine, Harry suddenly exclaimed that there was a detailed magnetic map of the area on which the ridge should be obvious. Fired up by that interjection, Fred rocketed out of the room and upstairs to the library, where grabbed volume 72 of the *Geological Society of America Bulletin* and slapped it down on Wilson's desk, open at pages 1268 and 1269 (Figure 11.3). All three stared in amazement at the left-hand page with its map. The magnetic anomalies paralleled the putative ridge, and there were symmetries about the crest. The map had been in the public domain for four years, yet no one had noticed the symmetry, which provided stunning support for seafloor spreading and reversals of the geomagnetic field.

However, with any big discovery there can be a devil in the detail, and there was in this case, too. When Vine superimposed the reversal timescale of Cox, Doell and Dalrymple on the map, he was dismayed to see that the pattern of marine magnetic stripes did not perfectly match the quantitative land-based reversal timescale of Cox *et al.* This was a showstopper because it implied implausible modulation (stops and starts) in the spreading rate over time. That was the main reason why the V–M hypothesis did not catch on immediately. In late 1965, Vine was in Kansas City making an effort to get traction for the V–M theory at the meeting of the GSA. He caught up with Dalrymple, who related to him that Doell and he were increasingly confident that they had discovered a new geomagnetic polarity event: it was a period of normal polarity lasting about 100,000 years located at about 0.9 Ma in the Matuyama reversed-polarity epoch. They had flagged it as the Jaramillo normal event, after Jaramillo Creek (Valles Caldera, Sandoval County, New Mexico), where they had found the evidence in six rhyolite domes with K–Ar ages of between 0.7 and 1.0 million years. Apprised of this news, Vine immediately realized it should be possible to interpret the Juan de Fuca sequence with a constant rate of spreading:

> To me, it was all over bar the shouting ... I could now make a
> compelling case that the sea floor spreads symmetrically about the
> mid-oceans but at an essentially constant rate ... it faithfully
> records the timescale of reversals of the Earth's magnetic field. ...
> The possibility of documenting the evolution of the present-day
> ocean basins and the geomagnetic reversal time scale for the past
> 150 million to 200 million years opened up.

Independent support for symmetrical seafloor spreading came quickly. Lamont scientists made an awesome discovery when they

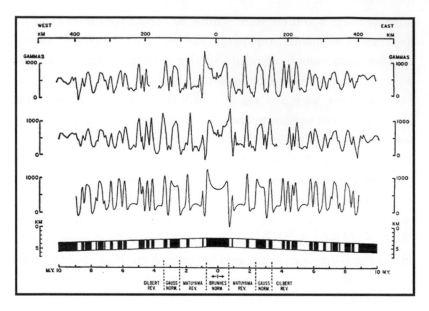

FIGURE 11.5 Presentation of the *Eltanin*-19 west–east magnetic anomaly profiles. The middle curve is the actual profile; east is to the right. The upper curve is *Eltanin*-19 reversed; west is now to the right. The lower curve is the model for the Pacific–Antarctic Ridge. The timescale (millions of years) relates to a spreading rate of 45 mm/year. The four magnetic epochs known at the time are indicated at the bottom of the figure.

Source: W. C. Pitman III and J. R. Heirtzler, *Science*, **154**, 1166 (1966). American Association for the Advancement of Science

examined the profiles shown in Figure 11.5 from cruises 19 and 20 of the US Navy Research Vessel *Eltanin*, which always towed a magnetometer in line with Maurice Ewing's infrangible instruction that his ships must be at sea harvesting all kinds of data every day. Another of his instructions was that such data must go into a common pool and be freely shared. When Walter Pitman processed magnetic data from *Eltanin*-19 and -20, he obtained a beautifully bisymmetric profile in which the magnetic anomalies fitted the entire reversal timescale known at the time (3.5 million years) and extended it to 10 million years. When Vine visited Lamont in March 1966 to see the *Eltanin* profile, he was at once convinced beyond doubt that the

V–M hypothesis was correct. Meanwhile, John Foster at Lamont got stuck into reversal stratigraphy of sediment cores, which yielded absolute ages. He, together with Lamont colleagues Neil D. Opdyke (a former student of Keith Runcorn) and Billy P. Glass, could thus connect magnetic reversals to dated samples from the seafloor rather than from the continents. This became the final confirmation of Hess's hypothesis of seafloor spreading, and it provided a massive breakthrough for Wegener's continental drift. It was a paradigm shift (or fundamental change of approach and underlying assumptions) with enormous implications for the future of Earth system science.

⊙ THE FINAL CHORUS

In October 1963, one month after the publication of the V–M hypothesis, Dan McKenzie became one of Teddy Bullard's graduate students, and so found himself in the same department of geophysics as Vine and Matthews. McKenzie attended Westminster School, founded in 1560, and the only ancient London school that is still on its original site, next to Westminster Abbey and the Houses of Parliament. This famous school has always emphasized liberal instruction and encouraged "loyal dissent"; that is to say, deep thinking and a willingness to challenge generally accepted ideas. He entered King's College, Cambridge, in 1960 to study natural sciences, but on finding the geology lectures "intellectually poor," he decided to concentrate on mathematics and physics. In his first year as a doctoral student, McKenzie followed Bullard's advice to work on atomic forces and thermodynamics with a view to producing an equation of state for the temperature and pressure of Earth's interior.

Sir Edward designed a research topic on Wegener's jigsaw puzzle for his doctoral student Jim E. Everett and research assistant Alan Smith (1937–2017), based on recovering the positions of the continents around the Atlantic before they drifted apart. By this stage, Bullard was a supporter of the V–M hypothesis, and his bright idea was to use Euler's fixed-point theorem of 1775, which states that any motion of a rigid body (such as a continent) on the surface of a sphere

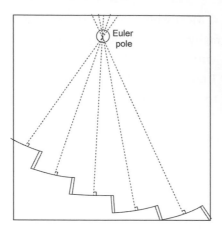

FIGURE 11.6 A transform fault is a horizontal shear fault along which two plates slip relative to each other. In 1965, three geophysicists at the University of Cambridge described geometrically the motion of a rigid body on a sphere in terms of rotation about a fixed point, the Euler pole. Great circles drawn perpendicular to the fault segments converge at the pole of rotation.
Source: Penny Wieser

(Earth) may be represented as a rotation of an appropriately chosen pole of rotation, as shown in Figure 11.6. The trio obtained the tightest fit (Figure 11.7) with the 500-fathom (915-metre) continental shelf contour. Their paper included the obligatory disclaimer that "a great deal of work needs to be done before we can fully accept the hypothesis that the Atlantic Ocean was formed by continental fragmentation."

Meanwhile, McKenzie worked on the problem of the equation of state, and he wrote up his results as a fellowship thesis that was awarded by King's College. Following his election to the Research Fellowship, he followed the liberal trajectory he had learned at Westminster School by pursuing his own agenda of academic inquiry on convection in the mantle. He did not get sidetracked by the hoopla because, as he later remarked, "[I]t was simply not obvious ... that what Fred and Drum were doing was so important." To investigate the underpinnings of mantle convection, McKenzie learned fluid mechanics and sufficient materials science to know that all materials creep at high temperatures and low stresses. By the time he had completed his thesis, he was convinced that the V–M hypothesis was indeed of major importance, and moreover that much work needed to be done on motion in three dimensions within the mantle. To accomplish that, he went west, to a postdoctoral position in the

FIGURE 11.7 The fit of Africa and South America using the 500-fathom (915-metre) contour. This geometrical fit, and similar analyses for other continental blocks, was achieved by representing plate motion as uniform around a Euler pole.
Source: E. Bullard, J. E. Everett and A. Gilbert Smith, "The fit of the continents around the Atlantic." *Philosophical Transactions of the Royal Society of London. Series A,* **258,** 41–51 (1965)

Seismological Laboratory at Caltech, Pasadena, where he finished the write-up of his PhD thesis and his first paper on plate tectonics. In this debut, he suggested that the ocean ridges are formed by upwelling of the hot mantle that everywhere lies below the spreading seafloor. McKenzie showed that this simple model for the seafloor, consisting of a slab of constant thickness moving with a constant velocity over the hot mantle, was able to account for the elevated heat flow of the spreading ridges. An attractive feature of this model was that it did not require a hot upwelling region of convection beneath the ridge: separation of the seafloor and upwelling of the mantle into the space thus vacated explained the origin of the ridges. No convection cells were required to keep the seafloor moving at a steady velocity on either side of the ridge. The elegance of this model lay in its use of analytic solutions to the equations.[20]

In April 1967, McKenzie attended his first meeting of the AGU in Washington, DC. After going through the abstracts, McKenzie decided to absent himself from the final talk of a session in which Jason Morgan of Princeton University was down to speak on a paper that was already published and that McKenzie had read. In the event, Morgan backed out of speaking on that paper – it was getting serious criticism – and instead pulled a rabbit out of the hat, promoting his latest paper titled "Rises, trenches, great faults, and crustal blocks."[21] Morgan had assumed that the great blocks (plates) would become rigid away from the rift, and he worked through the kinematics. But because this topic did not correspond with the abstract, the number of attendees was low, and those that were present mostly did not give the paper a second thought. McKenzie was busy elsewhere. And that's how matters stood for a few months.

McKenzie next ventured south to Scripps, La Jolla, in June 1967, where he eagerly read all of Hess's papers, because the seafloor spreading bandwagon was going full throttle on the West Coast. McKenzie teamed up with Robert (Bob) L. Parker, who had recently completed his doctorate in geomagnetism under Bullard and was now a postdoc at Scripps. Parker had written a map projection program that could be used to visualize the motion on a sphere. Following Bullard's lead, they worked on the fundamental principles of the motion of a rigid slab across a sphere: it was a "paving stone theory of world tectonics." They tested a model of seafloor spreading in which a slab (later terms a plate) interacted with another slab. They employed Euler's fixed-point theorem to handle the separate concepts of translation (horizontal motion) of a slab and its orientation (rotation to a new aspect) as uniform rotation about a single axis. For two (or more) slabs, the mathematical description of relative and indeed absolute motion then reduces to the motion of each slab about its pole of rotation.

Fred Vine had shown from the magnetic stripes on the seafloor that the spreading is at a uniform rate and that velocity can be turned into an angular rate of rotation about a pole. The next question was: How do you find the pole of rotation of a moving slab? This is where Tuzo Wilson's transform faults come into the story: the transform faults point at the slab's pole of rotation. By following the vectors to their point of convergence, the location of the rotation pole becomes known. But what McKenzie did not know was that Jason Morgan had already beaten him to it: that is, until a colleague returned from a conference at which Morgan had recounted the paper he had delivered to an inattentive audience at AGU. McKenzie realized in an instant that he and Morgan had had the same idea. Was it too late to write a paper? He turned to Bill Menard (1920–1986) at Scripps for advice. "Publish!" admonished the world authority on marine geology. Dan and Bob worked furiously for two or three days to cook up a paper for *Nature*. When they finished on a Saturday morning, they realized the post office at La Jolla had closed for the weekend. They fed quarters into a stamp machine until they had sufficient stamps to airmail the paper to London. *Nature* published it six weeks later.[22]

When McKenzie and Parker applied the mathematics of the paving stone theory to the North Pacific, which amounts to one quarter of Earth's surface area, they were astoundingly successful at explaining the tectonic setting and gross geology of the north Pacific Ocean, with its ridge, trenches, transform faults, active volcanoes and even the earthquakes triggered by gravity slides at the deep Kuril Islands trench. This powerful theory was the convergence of all of the research that had gone into continental drift, seafloor spreading, heat transmission from the mantle, magnetic anomalies, reversals of the geomagnetic field, gravity anomalies over the trenches, island arc volcanoes and so on. It was utterly transformative in that it created a new kind of science of Earth's interior. In the following chapters, we will explore how this enlightenment of geoscience has opened up the examination of carbon deep in Earth as a new kind of science.

REFERENCES

1. Olsen, P. L. *Biographical Memoir of John Verhoogen 1912–1993* (National Academy of Sciences, 2011).
2. Clegg, J. A., Deutsch, E. R. and Griffiths, D. H. Rock magnetism in India. *Philosophical Magazine* **1**, 419–431 (1956).

3. Cox, A. and Doell, R. R. Review of palaeomagnetism. *Geological Society of America Bulletin* **71**, 645–768 (1960).

4. Frankel, H. Hess papers, Princeton University. Letter to Vening Meinesz dated 6 July 1959. Harry Hess develops seafloor spreading. In *The Continental Drift Controversy* (Cambridge University Press, 2012), pp. 198–279.

5. Cox, A., Doell, R. R. and Dalrymple, B. G. Reversals of the Earth's magnetic field. *Science* **144**, 1537–1543 (1964).

6. Mason, R. Stripes on the sea floor. In *Plate Tectonics: An Insider's History of the Modern Theory of the Earth* (Westview Press, 2001), pp. 31–45.

7. Menard, H. W. *The Ocean of Truth: A Personal History of Global Tectonics* (Princeton University Press, 2014).

8. Mason, R. G. and Raff, A. D. Magnetic survey off the West Coast of North America, 32° N. latitude to 42° N. latitude. *Geological Society of America Bulletin* **72**, 1259–1265 (1961).

9. Raff, A. D. and Mason, R. G. Magnetic survey off the West Coast of North America, 40° N. latitude to 52° N. latitude. *Geological Society of America Bulletin* **72**, 1267–1270 (1961).

10. Morley, L. W. The zebra pattern. In *Plate Tectonics: An Insider's History of the Modern Theory of the Earth* (Westview Press, 2001), pp. 67–85.

11. Gratzer, W. John Maddox (1925–2009). *Nature* **458**, 983–984 (2009).

12. Vine, F. J. Reversals of fortune. In *Plate Tectonics: An Insider's History of the Modern Theory of the Earth* (Westview Press, 2001), pp. 46–66.

13. Dietz, R. S. Continent and ocean basin evolution by spreading of the sea floor. *Nature* **190**, 854–857 (1961).

14. Vine, F. J. and Matthews, D. H. Magnetic anomalies over oceanic ridges. *Nature* **199**, 947–949 (1963).

15. Talwani, M. A review of marine geophysics. *Marine Geology* **2**, 29–80 (1964).

16. Hess, H. Seismic anisotropy of the uppermost mantle under oceans. *Nature* **203**, 629–631 (1964).

17. Matthews, D. H., Vine, F. J. and Cann, J. R. Geology of an area of the Carlsberg Ridge, Indian Ocean. *Geological Society of America Bulletin* **76**, 675–682 (1965).

18. Cann, J. R. and Vine, F. J. A discussion concerning the floor of the northwest Indian Ocean – an area on the crest of the Carlsberg Ridge: petrology and magnetic survey. *Philosophical Transactions of the Royal Society of London. Series A, Mathematical and Physical Sciences* **259**, 198–217 (1966).

19. Wilson, J. T. A new class of faults and their bearing on continental drift. *Nature* **207**, 343–347 (1965).

20. McKenzie, D. P. Some remarks on heat flow and gravity anomalies. *Journal of Geophysical Research* **72**, 6261–6273 (1967).

21. Morgan, W. J. Rises, trenches, great faults, and crustal blocks. *Journal of Geophysical Research* **73**, 1959–1982 (1968).

22. McKenzie, D. P. and Parker, R. L. North Pacific: an example of tectonics on a sphere. *Nature* **216**, 1276–1280 (1967).

12 Deep Carbon

Cycles, Reservoirs and Fluxes

◉ THE JIGSAW SLOTS TOGETHER

This chapter completes the story of the acceptance of plate tectonics, which marks the beginning of the modern period of Earth system science. This final approval required additional contributions by several researchers, with the key papers being published in 1968. Jason Morgan's work on crustal blocks, of which he had given an impromptu preview at the American Geophysical Union in April 1967, proposed that on Earth's dynamic surface 20 crustal blocks move relative to each other, endlessly jostling for their place in the jigsaw. Simultaneously, Xavier Le Pichon, a young French geophysicist who worked with the Ewing brothers at Lamont Geological Observatory from 1963 to 1989, connected the kinematic ideas of Morgan, McKenzie and Parker to the vast data sets held at Lamont, particularly the magnetic profiles. Le Pichon's computer model demonstrated that the motion of six large rigid blocks completed a jigsaw that covered most of Earth and could accurately account for the evolution of ocean basins. His model of June 1968 indicated that plates do indeed form an integrated system in which the sum of all crust generated along 50,000 kilometres of ocean ridges equals the cumulative amount destroyed in the subduction zones.[1]

In September 1968, Jack Oliver (1923–2011), Bryan Isacks and Lynn R. Sykes of Lamont further refined Le Pichon's model by bringing seismic data into play.[2] Their global plot of 30,000 earthquake epicenters (Figure 12.1) revealed the outlines of the major lithospheric plates. There are shallow-focus earthquakes at the ridges and deep-focus earthquakes at the trenches. They introduced the new term "global tectonics," and their iconic cartoon (Figure 12.1) shows slabs

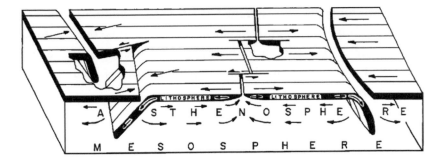

FIGURE 12.1 This simple block diagram illustrated the roles of the lithosphere, asthenosphere and mantle (mesosphere) in the new global tectonics. At the left, there is a transform fault between two oppositely facing zones of convergence. In the center, there are two ridge-to-ridge transform faults.
Source: Bryan Isacks, Jack Oliver and Lynn Sykes. Seismology and the new global tectonics. *Journal of Geophysical Research* **73**, 5857 (1968)

descending into the asthenosphere, where the pattern of circulation is opposite to the motion of the overlying plates. This model suggested that in the subduction zone gravitational forces pull the slab vertically down toward the mantle, and on the ridges the same force pushes the plates away horizontally. It was exceptionally helpful during the early development of the physical theory of plate tectonics.

A key point is that the model was *not* about the lithosphere being forced into a geometric fit with convection cells in an idealized model of Earth. Rather, the downward motion of the cold (and heavy) slab was attributed to its gravitational potential energy being released as it sank into the abyss, rather than the push of thermal energy from a convection cell. The final paragraph of their 1968 paper has this to say:

> The worldwide phenomena of seismology provide crucial evidence on the basic processes of the earth's interior that have shaped and are shaping the surficial features of interest to classical geology. Even if it is destined for discard at some time in the future, the new global tectonics is certain to have a healthy, stimulating, and unifying effect on all the earth sciences.

This powerful prediction launched a geological revolution, stimulating geoscientists to network in global terms and to be open to fresh ideas. By the 2010s, progress in the theory led to a holistic picture that integrated the interplay between plate tectonics and mantle convection.[3]

⊙ RING OF FIRE

An American naval officer, Commodore Matthew Calbraith Perry (1794–1858), first drew attention to the horseshoe of volcanic activity surrounding the Pacific Ocean in his official report to the US Congress of the Naval Expedition to East Asia. In the introduction, Perry mentions "the immense circle of volcanic development which surrounds the shores of the Pacific from Tierra del Fuego around to the Moluccas."[4] The kingdom of Japan comprises "a great number of islands, said to number 3850" that can be traced as far as Alaska "on our own continent." The commodore warns that the volcanic islands have coasts that are difficult to access, with sunken rocks, dangerous whirlpools and violent winds. An almost continuous sequence of deep oceanic trenches, volcanic island arcs and volcanic belts extending for 40,000 kilometres rings the Pacific basin and is the locus of about 90 percent of our planet's earthquakes and three-quarters of its active and dormant volcanos.

In December 1969, a highly significant Geological Society of America (GSA) conference assembled at Asimolar in Pacific Grove, California. William R. Dickinson (1931–2015), a geoscientist at Stanford University, had proposed and then organized this week-long gathering of experts in order to find a new synthesis for "the earlier ideas of continental drift, seafloor spreading, and the deep structure of island arcs" within the framework of the movement of "large, semirigid plates of crust and upper mantle." This was the world's first GSA Penrose Conference, to which any geoscientist in the world could seek invitation to its "open and frank discussions in an informal setting and to stimulate individual and collaborative research." About 150 geoscientists attended, and according to Dickinson,

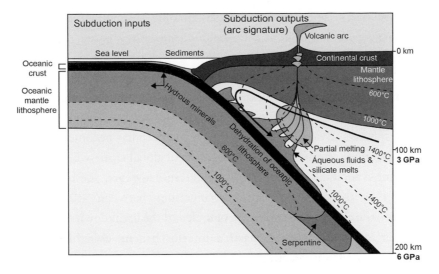

FIGURE 12.2 A modern rendering of the principal "moving parts" involved in the tectonics of Earth's interior. To the right of the diagram, the units indicate depth and pressure.
Source: Penny Wieser

"[O]ur discussions continued day and night at a furious tempo that left us all exhausted."[5]

This gathering enthusiastically adopted the term "subduction zone" to describe the leading downward of Earth materials by the descent of a tectonic plate from the surface of Earth back into the interior. Diagrams similar to Figure 12.2 then began to appear in the literature to show an oceanic plate of lithosphere, composed of rigid crust and upper mantle, descending at an angle into the hotter and deeper asthenosphere. The subduction zone marks the locus of plate descent, and the magmatic arc forms above it as molten rock rises upward from the mantle into the crust, causing uplift and volcanism. In the majority of cases, the slab does not melt. What happens is that intense metamorphism dehydrates minerals, releasing fluids that reduce the temperature of the mantle wedge above. The cross-section in Figure 12.2 shows the principal "moving parts" of the deep cycle that lead to burial of carbon at the subduction zone in the form

of carbonates, sediments and fluids and its subsequent return by
outgassing and eruption at the volcanic zone.

⊙ INVENTING VOLCANOLOGY

In modern times, understanding the processes and consequences by
which deep carbon has been, and continues to be, transferred from the
atmosphere and oceans to reservoirs in the deep has become an enormous frontier of research. The mitigation of global climate change has
become a societal and political imperative. However, the basic concept that there are chemical connections between atmospheric
carbon, the chemical weathering of rock and the breakdown of organic
matter and carbonates in Earth's interior has its own history. In
Chapter 6, I summarized Lavoisier's pioneering research on respiration to illustrate that the concept of a deep carbon cycle linking
carbon in Earth's interior with carbon in the air is not new. Daring
field research conducted by continental geoscientists in the eighteenth and nineteenth centuries features prominently in this history.[6]

In the eighteenth century, Italy was the only *accessible* region
in Europe with active volcanoes, so that's where I'll begin the history
of volcanology and carbon transfer to the atmosphere. When the Abbé
Lazarro Spallanzani (1729–1799) held the chair of natural history at
the University of Pavia, he was an indefatigable traveler in pursuit of
geological specimens for the public Museum of Natural History in
Pavia. It took him only 10 years to transform its mediocre collection
into the most magnificent in Italy.[7] In 1788, he sauntered to the
Kingdom of the Two Sicilies with his geological hammers and collection bags, hoping to augment the Museum's volcanic specimens.
Perilous field trips to Vesuvius, Stromboli, Vulcano and Etna were
all part of the fun for this bold pioneer. One of his sorties to the
summit of Vesuvius ceased abruptly with a violent eruption, but
before scurrying away he valiantly edged himself to within 2 metres
of the lava flow to record its velocity. On another foray, when just
short of the crater rim of Etna, he collapsed unconscious after choking
on its toxic gases. Nevertheless, he later nonchalantly returned to

experience the "pleasure" of peering down at the turbulent lava. On Stromboli, he made the highly significant observation that the powerful venting of gas was forcing the lava to ascend and explosively eject enormous rocks, which he avoided by cowering in a cavern. Spallanzani's attempts to collect samples of the gases emanating from the crater of the volcano Stromboli were not successful.[8] Undeterred by these near-death experiences, he gingerly stepped down into the crater of Vulcano, only beating a hasty retreat when his wooden staff caught fire and his feet were scorching hot. It was all in the pursuit of science. He issued the following warning to future students of volcanoes:

> That these and other similar experiments are difficult, offensive, and, in some degree even dangerous; but what experiment can be undertaken perfectly free from inconvenience, and all fear of danger, on mountains which vomit fire? I would certainly advise the philosopher who wishes always to make his observations entirely at his ease, and without risk, never to visit volcanoes.[9]

He consolidated his findings from this heroic fieldwork with high-class laboratory analysis. Being already familiar with the glass furnaces of Murano, Venice, he built his own furnace, together with a pyrometer invented by the English potter Josiah Wedgwood for measuring the temperature of clay being fired in a kiln.[10] Thus it was that Spallanzani became the first geoscientist to apply quantitative experimental analysis to the investigation of volcanic rocks. By melting lava in his furnace and gaging the melting point with a Wedgwood pyrometer, he determined the composition and fusion temperature of lava, as well the gases expelled. Spallanzani was on the right track to the foundation of a new science, even though his published results were not conclusive.

Another turning point had arrived in 1774. After centuries of neglect by the natural philosophers, Sir William Hamilton (1730–1803), the British Ambassador in Naples (back then the capital of a large, independent southern kingdom under the Spanish

Bourbons), quietly initiated volcanology by publishing his personal observations of eruptions. These eye-witness accounts provided information and a stimulus for Spallanzani, by then the most accomplished natural philosopher in Italy, to develop a theory of volcanic activity in terms Newton's physics and Lavoisier's chemistry. His most remarkable suggestion was that the release of water or carbonic acid from minerals under pressure sustained the explosive volcanic gas emissions, and thus led naturally to the restitution of those volatiles in the atmosphere. Spallanzani's observations, experiments and publications position him today as a key contributor to the early history of research on the deep carbon cycle.

In London, the chemist Humphrey Davy (1778–1829) pointed out that "our existence itself depends on the equilibrium of gases in the atmosphere," with the oxygen consumed by respiration and combustion being replaced somehow. By then, several botanists and chemists had already demonstrated that green plants purify the air. In 1799, Davy described how volatiles are recycled through a "perpetual series of changes [on which] life appears to depend," and therefore:

> The vegetation of land plants then may be considered as the great source of the renovation of the atmosphere of the land animals. . . . Analogy led me to suppose that sea vegetables must be the preservers of the atmosphere of the oceans.[11]

Nicolas-Théodore de Saussure (1767–1845) was a Swiss chemist and plant physiologist who extended Davy's work by rigorously demonstrating that water is decomposed during plant growth and oxygen is released: he correctly outlined the chemical transformations that take place during photosynthesis. The next generation of European geologists recognized de Saussure as a pioneer in identifying the global "rotation of carbon" as a law of nature. He effectively laid the groundwork for developing a global perspective of carbon cycling via reservoirs and fluxes.[12]

Our narrative now takes us to Jean-Baptiste Boussingault (1802–1887), an adventurous French chemist from the École des

Mines at Saint-Étienne and the innovator of the chemical analysis of volcanic outgassing. General Simón Bolívar, the famous military and political leader who had liberated the Spanish colonies that became the republics of Venezuela, Bolivia (named for him), Colombia, Ecuador, Peru and Panama, appointed Boussingault as his mining engineer in Venezuela, with the rank of colonel. Over a period of 10 years, Boussingault traveled widely in the northern part of the continent, making a picaresque progress in the mining regions while avoiding rebellious soldiers and unreliable guides, as well as coping with the assassination in Quito of a military colleague. When Bolívar died of tuberculosis in December 1830, Boussingault had already become disenchanted with the chaotic political situation in the state of Gran Colombia. But before departing for France, he decided to augment his memories of the extinct volcanoes in his native Auverne with some live action in the Andes. He already had experience of climbing volcanoes and investigating their exhalations, but now his ambition stretched to some of the highest volcanoes in Colombia and Ecuador. Between March and December 1831, he attempted to climb seven volcanoes. He reached the summits of Cotopaxi (5986 m) and Cumbal (4764 m), the latter of which greatly excited him because it was partially active: "[T]he scene was magical, flames emerging from a mass of snow." But Chimborazo (6310 m), then considered the world's highest mountain, and unconquered at the time, defeated him on account of falling snow and his respiratory difficulties at high altitude.

On these ascents, Boussingault carried out fieldwork under challenging conditions. With his small portable laboratory for analyzing minerals, as well as jars for specimens of gases and snow, he braved the scorching vapors of fumaroles to investigate whether any of the observable phenomena were temperature dependent. The compositions of gases (CO_2, H_2S and O_2) were measured directly in the field. Carbonic acid gas (CO_2) accounted for over 90 percent by volume of the samples.[13] When he published these results in 1832, he concluded that the gas and water have "their source in the volcanic furnaces." Furthermore, he suggested that thermal instability of carbonates at depth would produce reduced carbon. Two years later, in a treatise on the composition of the atmosphere, he recognized the chemical action of plants on the atmosphere, noting that atmospheric carbon dioxide is the source of carbon for all living beings on a global scale:

> ... many causes tend to annually modify the composition of the atmosphere ... one of them occurring at the surface of Earth due to life itself. Organised beings are taking part of the carbon that is in the air ... In plants this assimilation occurs directly ... [but in animals this assimilation] is by eating plants.[14]

In summary, Boussingault achieved a magnificent synthesis through his recognition of the role of plants and of the significance of the burial and preservation of organic deep carbon in sediments for the long-term composition of the global atmosphere. He was the pioneer who first understood that volcanic gases need to be included in the fluxes of carbon venting into the atmosphere *"sans interruption."*

◉ JACQUES-JOSEPH ÉBELMEN: DEEP CARBON CYCLE PIONEER

By the early nineteenth century, French savants had a global vision of the circulation of carbon on Earth and within Earth thanks to discoveries of the sources and sinks of atmospheric carbon dioxide. In 1828, botanist Adolphe Brongniard, who was keen that geologists should make use of botany, recognized that vegetation had a global purifying effect. He reasoned that the slow accumulation of deep carbon in the form of coal, lignite and bitumen in sedimentary strata implied that the atmosphere had contained more carbon dioxide in the past. By the 1830s, it was accepted in France that the burial of carbonaceous material in sediments was a major sink of atmospheric carbon dioxide.

Meanwhile, chemists and geologists across Europe latched onto the dissolving power of carbonic acid in water–rock reactions. Karl

Gustav Bischof (1792–1870) at the University of Bonn brought chemical and physical analysis into general use in geology through an enormous textbook. The first volume is devoted to the chemical weathering of rocks by carbonic acid in water and the consequential release of soluble elements through the progressive decomposition of silicate rocks.[15] Bischof effectively contributed to the discovery of chemical weathering as we would describe it today. His younger contemporary, the brilliant French mineralogist Jacques-Joseph Ébelmen (1814–1852), extended the study of chemical weathering in three major papers, published shortly before his untimely passing at the age of 37.[16–18]

Ébelmen enrolled at the École Polytechnique in 1831. The elite school then operated as a military academy. He spent two formation years there, before moving to the École des Mines in 1833, where he graduated with the highest honors, as well as a deep grasp of chemistry and mineralogy that was to secure his career. His first assignment as a government mining engineer was at Vesoul, where ore deposits of economic importance required investigation. His soaring reputation soon led to a recall to Paris in 1841 to be Secretary for the *Annales des Mines*, one of the world's oldest journals (established in 1794) for promoting applied science and technology; the *Annales* continues to be edited under the ministerial authority of the French state. At the end of 1845, Ébelmen won the appointment of chief engineer of mines at the Sèvres national porcelain manufactory, as well as the chair of mineral assay at the École des Mines. At Sèvres, Ébelmen improved the quality of the porcelain as a result of carrying out detailed chemical analysis of the weathering of basalt, the source rock for the formation of kaolinite and petuntse. He found that carbonic acid causes alkalis, alkaline earths and silica to get lost in solution, leaving a solid residue. Having made these *chemical* investigations of kaolin and other materials for commercial exploitation, Ébelmen subsequently put his discoveries into a *geological* context that encompassed the science of a deep carbon cycle.

His research showed that the chemical action of acidic rain-water leads to the dissolution of calcium carbonate rock and the formation of caves. Ébelmen also surmised correctly that the roots of land plants are vital for speeding up the weathering of rock. His treatise of 1845 includes a masterly summary of the processes governing the changing levels of carbon dioxide and oxygen in the atmosphere over geological time. In addition to the role of carbonic acid on silicates as a sink for atmospheric carbon, Ébelmen pointed out that carbonate precipitation in the oceans through the medium of zoophytes or mollusks would constitute another carbon sink. To counterbalance the effects of the absorption of carbon dioxide through the decomposition of rocks, he turned to Boussingault's notion that volcanic gases are the source of atmospheric carbon dioxide, although he did not consider the ultimate origin of the gases emitted by volcanoes, an issue he preferred to leave as an open question for the next generation. Incidentally, in his papers, Ébelmen did not use words alone: he wrote down the reactions in chemical notation that we can easily recognize, a feat that is unique in the geological litera-ture of the mid-nineteenth century.

He published his conclusions in the *Annales des Mines*, where they remained unnoticed for 35 years until being lauded by the American chemist Thomas Sterry Hunt (1826–1892). A graduate in chemistry from Yale University, Hunt had been hired by the Geological Survey of Canada as a chemist and mineralogist in 1847. For the next 25 years, he lived and worked in Montreal, where he mingled with French Canadian polite society, conversed in perfect French and, more significantly, had ready access to the French scien-tific literature. In 1863, he published a short article on the history of Earth's climate, in which he argued that carbon in the forms of limestone and mineral coal must once have been in the atmosphere as carbon dioxide, where it acted as a greenhouse gas, contributing "much to the mild climate of the Carboniferous age."[19] Hunt con-tinued to work on the connection between chemical geology and the atmosphere, and then, in the late 1870s, he came across Ébelmen's papers in *Annales des Mines*. After studying them, Hunt became deeply impressed by Ébelmen's conclusion that the geochemical weathering of rock could inform us about the composition of the atmosphere in ages past. In August 1878, Hunt publicized Ébelmen's achievements at the Dublin meeting of the British Association for the Advancement of Science. Hunt also tried to bring the matter to the attention for the French Academy of Sciences.[20] Unfortunately for Ébelmen's reputation – by then he had been dead for 26 years – Hunt's thoughtful initiatives were of no avail.

Two silent decades on the carbon cycle passed until 1894, when the Swedish chemist Arvid Gustaf Högbom (1857–1940) of Uppsala University considered the effect on the atmosphere and climate that would result from long-term changes in the inputs and outputs of the global carbon cycle. He examined nine different pathways for the inputs and outputs of atmospheric carbon dioxide. In the case of the deep carbon reservoirs, he inferred that enormous amounts of carbon must have become buried in Earth's crust as calcium carbonate, which led him to surmise that the creation of the deep carbon reservoir must have been balanced by volcanic outgassing: otherwise, the proportion of carbon dioxide in the atmosphere would be unstable over geological time.[21] Continuing with the theme of atmospheric carbon dioxide, two seminal papers on the greenhouse effect were soon published: one by Svante Arrhenius in 1896 on the role of carbon dioxide in the heat balance of Earth, and the second by T. C. Chamberlin three years later.[22,23] Arrhenius quoted extensively from Högbom's paper of 1894, for which he provided excellent English translations. The ironic twist of history is that Arrehenius's translation of Högbom's work was so helpful that later workers misattributed the credit to the translator rather than the originator! Chamberlin's paper of 1899 was heavily based on Högbom's work, copiously copied without proper citations.[24]

⊙ BREATHING NEW LIFE INTO DEEP CARBON SCIENCE

In the second half of the twentieth century, the American physical chemist Harold Urey (1893–1981) deservedly received much credit for formulating the silicate–carbonate subcycle. His early career research in isotope chemistry won him the Nobel Prize in Chemistry in 1934 for the discovery of deuterium. By 1939, Urey and his collaborators had developed mass spectrometers that were sufficiently sensitive to investigate the rare isotopes of carbon, nitrogen and oxygen.

Isotopes of light elements can be used to trace the rates of change of organic carbon stored in the crust and mantle over long timescales. First, though, a reminder that carbon on Earth occurs as two stable isotopes: 98.9 percent in the form of ^{12}C and 1.1 percent in the form of ^{13}C. One of several pathways for transferring carbon from the atmosphere to the interior is the chemical reaction of photosynthesis, used by plants to convert light energy from the Sun into chemical energy that can fuel the organisms' life cycle. In the photosynthesis reaction, metabolic processes take up ^{12}C in preference to ^{13}C, which leads to the organic matter derived from organisms, such as marine carbonates, fossils or humus, being depleted in ^{13}C. Measurements of the deficiency in ^{13}C across geochemical or geological samples of different ages are essential for modeling the carbon cycle and its changes through geological time. The basis of the technique is to measure the $^{13}C:^{12}C$ ratio in a sample and compare it with an agreed standard, Vienna Pee Dee Belemnite, for which the ratio is 0.01118. The amount by which a sample differs from the standard is reported in parts per thousand (per mil, ‰), known as delta carbon and written with the symbol "$\delta^{13}C$." Most natural materials have a negative $\delta^{13}C$.

By early 1949, Urey, by then at the University of Chicago, had built high-precision mass spectrometers suitable for the isotopic analysis of light elements. He became the "world expert" in this new field of geochemistry and trained a succession of doctoral students. His research team applied techniques of carbon and oxygen isotope geochemistry to establish a paleotemperature scale for calcium carbonate precipitated from water. Through an extraordinarily imaginative analysis of $^{18}O/^{16}O$ fractionation, they uncovered the life history of a belemnite fossil (Cretaceous period, 100 Ma) from the Isle of Skye: it had lived through four summers (mean temperature 21°C) and three winters (15°C). Three years later, one of Urey's doctoral students, Harmon Craig (1926–2003), published a far-reaching paper that established him as the founder of carbon isotope geochemistry.[25] He established the $\delta^{13}C$ characteristics of carbon in meteorites, the oceans and their sediments, the atmosphere and plants, as well as deep carbon in igneous rocks, diamonds and volcanic gases – and that was just the fieldwork and laboratory analysis for his doctoral thesis.

In the later stages of a stellar academic career, Urey became interested in space science, the origin of life and the chemistry of Earth's atmosphere. In 1952, the latter topic led him to propose the following equilibrium reaction for calcium carbonate and, to a lesser extent, magnesium carbonate:

$$CO_2 + CaSiO_3 \Leftrightarrow CaCO_3 + SiO_2$$

$$CO_2 + MgSiO_3 \Leftrightarrow MgCO_3 + SiO_2$$

carbon dioxide + calcium (magnesium) silicate $\Leftrightarrow$
calcium (magnesium) carbonate + silicon dioxide

This is now recognized as the Ébelmen–Urey reaction.[26] Reading from left to right, the reaction represents removal of carbon dioxide via weathering by Earth's humid atmosphere and carbonate precipitation in the oceans. From right to left, it represents the release of carbon dioxide when carbonates and silicates undergo metamorphic change; for example, as a result of subduction. The carbon dioxide from this reaction escapes through fissures and cracks in the crust. The important steps in the reaction when read from right to left had already been verbally articulated in Ébelmen's paper, which was more than a century ahead of its time. In the course of his employment as an industrial chemist, Ébelmen correctly deduced the long-term processes affecting the exchange of CO_2 and O_2 between rocks and the atmosphere. Such processes included not just volcanism, but also the role of plants in weathering, the weathering and burial of organic matter and pyrite and the weathering of basalt. Ébelmen's achievements and foresight remained buried deep in the literature until 1996. That's when Robert A. Berner (1935–2015), professor of geology and geophysics at Yale University, sounded a wake-up call for geoscientists to celebrate Ébelmen's achievements

> ... as a truly major figure in the history of Earth science [who] laid a firm foundation for much of the present-day discussion of global change over geological time.[27]

⊙ BOX MODELS OF THE LONG-TERM CARBON CYCLE

The pathfinders of deep carbon science who were introduced above were hampered in their theoretical development by a lack of observational and experimental data. Berner demonstrated in 2012 that Ébelmen's concepts are consistent with what we now know about how to model the deep carbon cycle, which is instructive from the point of view of the philosophy of science.[28] Concepts such as data sharing, mathematical analysis and computation lay in the future for Ébelmen, who could only describe and hypothesize. Our journey through the history of the geosciences has thrown up examples where a new approach to a problem did not catch on: the investigative tools to falsify or to support a hypothesis simply did not exist. Ébelmen got the chemistry correct. Urey's data were superior, his chemical analysis informative and he did simple modeling of the carbon cycle over time by using reaction rates.

A significant move forward with the theory of the Earth system was the development of box models for the exchange of carbon in nature between its geochemical reservoirs. In models of this kind (Figure 12.3), each reservoir is represented by a box that is connected to others by pipes. This is mathematical modeling in which the natural carbon cycle is taken to be a chemical factory, with boxes

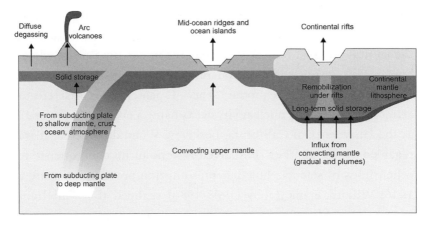

FIGURE 12.3 Sketch of the storage reservoirs and fluxes for carbon subducted to the deep mantle. Carbon dioxide is returned to the atmosphere through degassing of the ocean, arc volcanoes, mid-ocean ridges and ocean islands.
Source: Penny Wieser

holding the carbon dioxide before and after transfer and fluxes controlled by pipes, pumps and valves. The short-term (seasonal, annual, centuries) and long-term (millions of years) features of the system are investigated using time-dependent differential equations. Scientists building such models choose the initial conditions and various parameters, and then the geodynamic properties of the situation are handled through mathematical analysis (1950s and 1960s) or numerical coding (from the 1970s).

The first feasible models of Earth's interior appear in the late 1950s. Numerical models to reconnoiter the exchange of carbon dioxide between reservoirs over the geological timescale date from 1957. That's when Harmon Craig, Roger Revelle and Hans Suess (1909–1993) of the Scripps Institution of Oceanography made quantitative estimates of the absorption and exchange rates of CO_2 isotopes between the atmosphere and the oceans. Two papers from Scripps in 1957 are of historical importance to the deep carbon story because they are the first attempts to estimate the average residence time of carbon dioxide in the atmosphere before its take-up by the ocean.

Craig used carbon isotope ratios to investigate the natural distributions of ^{12}C, ^{13}C and radioactive ^{14}C. (As an aside, I should add that the natural origin of ^{14}C takes place in the upper atmosphere when a high-energy cosmic ray collision releases neutrons that react with ^{14}N to form ^{14}C and a proton.) The ^{14}C isotope was first created artificially in 1940 by nuclear physicists using a cyclotron accelerator at the University of California, Berkeley. When discovered, it was regarded as yet another short-lived isotope of no consequence for other fields of science, but that presumption proved premature in 1946: Willard Frank Libby (1908–1980) realized that "all living matter, [and] dissolved matter in the oceans" would naturally include ^{14}C because of its half-life, which he determined to be 5000 years.[29,30]

> The discovery of cosmic-ray carbon has a number of interesting implications in the biological, geological, and meteorological fields; a number of these are being explored, particularly the determination of ages of various carbonaceous materials in the range of 1,000–30,000 years.[30]

The determination of ^{14}C concentrations in the oceans and atmosphere opened up the possibility of finding the rates of exchange and turnover in the short-term carbon cycle.[31] Revelle and Suess examined whether the CO_2 flux from the exponential rise of industrial fuel production could account for "the observed slight rise of average temperature in northern latitudes during recent decades."[32] Both put the residence time of CO_2 in the atmosphere at 10 years, as did the Swedish meteorologist Bert Bolin (1925–2007), who would serve as the first chairman of the Intergovernmental Panel on Climate Change (IPCC) from 1988 to 1997. Although a lack of sufficient data prevented Revelle and Suess from foreseeing future changes in atmospheric CO_2, they did recommend that the forthcoming International Geophysical Year (IGY) would provide an opportunity "to obtain much of the necessary information" through global cooperation. Hence, from 1957, the application of box diffusion models to recover the history of the global carbon cycle over geological time became a staple of geodynamics and geochemistry.

Systematic measurements of the concentration of atmospheric CO_2 were initiated by Charles David Keeling (1928–2005). After a three-year postdoc position in geochemistry at Caltech from 1953 to 1956, he became affiliated with Scripps, where he worked for the rest of his life. Keeling developed an accurate system for monitoring the concentration of CO_2 in the background air, and with Revelle's support, he began extensive air-borne and ship-borne measurements. He received IGY funding to record CO_2 levels at a station on Mauna Loa in Hawaii at an altitude of 3000 metres and far from land-based sources of CO_2. In 1960, he established that strong seasonal variations of CO_2 were due to plant growth in the summer causing atmospheric CO_2 levels to fall, followed by a strong peak in late winter. The following year, his data were showing a steady rise of CO_2. From then until his passing in 2005, the concentration of CO_2 increased from 0.0315 percent to 0.0380 percent, a reflection of the staggering rate at which deep carbon was and continues to be removed from its subterranean reservoirs and returned to the surface reservoirs. By the 1970s, Bolin had become the world authority in this area for his quantitative estimates of the effects of human activity on the relentless rise in atmospheric CO_2 (shown in Figure 12.4) as revealed in the data from Mauna Loa and elsewhere.

In the early 1970s, box diffusion models for the terrestrial carbon cycle became more sophisticated with the introduction of features such as mixing, eddy diffusion and time-dependent parameters for tracking changes in reservoir content and fluxes. In 1974, four physicists at the Physics Institute of the University of Bern, Switzerland, produced a model that could not be analyzed by mathematical functions. They appear to have been the first group to make the move from solving partial differential equations by hand to Fortran programming on an IBM/370 computer. To handle time dependencies, they used a linear chain of 43 boxes and calculated the fluxes from box to box. They pronounced their digital model as

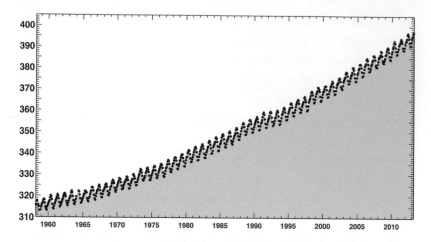

FIGURE 12.4 The monthly mean carbon dioxide (1958–2013) at the Mauna Loa Atmospheric Baseline Observatory, Hawaii. The units are parts per million (ppm). In May 2019, Mauna Loa data established a record high of 414.7 ppm. The upward trend continues to accelerate. Mauna Loa data are collected at an altitude of 3400 metres in the northern subtropics.
Source: National Oceanic and Atmospheric Administration

"well suited for the simulation of non-steady state phenomena using real observed input data."

At the height of his career, Berner focused on the long-term carbon cycle.[33] He became a leading proponent for modeling of the deep carbon cycle in terms of changes in the reservoirs and fluxes over geological time, as opposed to the recent past. In Earth's interior, carbon exists in several mineral phases, according to temperature and pressure. Magma from the upper mantle lifts deep carbon to the surface reservoirs in the form of carbon dioxide, diamond, graphite and carbides. The seafloor conveyor belt transfers carbon-bearing sediments from the surface to the interior through subduction. The Earth system moves carbon in many forms: minerals, living and dead organic material, carbon dioxide and methane and carbon dissolved in fluids such as silicate melts. These materials may alter as they move from one reservoir to another: water and carbon dioxide in mantle rocks can reduce rock viscosity by an order of magnitude. One of the key issues for long-term modeling is to build quantitative knowledge of the location, extent and nature of the reservoirs in the mantle and their rates of change over geological timescales. In 1989, Berner and the geochemist Antonio C. Lasaga, also of Yale, declared:

> The most vexing problem we have encountered in modelling the geochemical carbon cycle is to calculate the rate of degassing of carbon dioxide due to igneous and metamorphic activity. ... With our computer model we have shown that the primary factor affecting the carbon dioxide level is the rate of degassing. ... The world climate in general [is] controlled by tectonism, processes taking place deep within the Earth.[34]

Berner was a groundbreaking theorist who incorporated into his modeling many complex feedbacks. These linked carbon burial due to subduction into the mantle with subsequent outgassing, nutrient cycling and atmospheric carbon dioxide and oxygen.[35] He led the way in connecting the ocean–atmosphere carbon cycle with the geochemical carbon cycle of rocks and the deep Earth. To do so, Berner simplified mathematical manipulation with a fairly simple box model, and that freed him to increase the geological and biological complexity with great effect. His model had time steps of one million years for estimating the variations in the level of atmospheric carbon dioxide over the past 570 million years. In this way, his model replicated the warming of the global climate during the Mesozoic (240–65 Ma) and early Paleozoic (570–350 Ma) and the low levels of warming during the late Paleozoic (330–260 Ma) and late Cenozoic (last 30 million years).[36] By the close of the twentieth century, global climate science had accelerated the introduction of models of increasing complexity at scale with vast improvements in information technology (Figure 12.5). Although most of this science was principally concerned with climate change, Berner continued to develop his modeling of the deep carbon cycle to establish how the long-term cycle has affected the burial of organic carbon, fossil fuel formation and the evolution of the atmosphere over Phanerozoic time.[37]

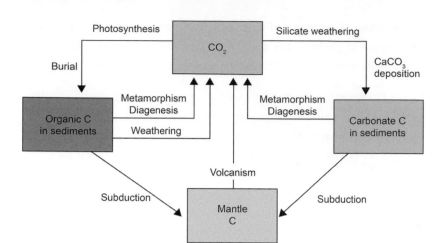

FIGURE 12.5 Box model of the long-term carbon cycle occurring over millions of years. It is subdivided into two cycles: one involving organic matter (left-hand side) and the other involving carbonate deposition (right-hand side). Carbonate deposition derived from carbonate weathering leads to degassing of carbon dioxide on deep burial and thermal decomposition. *Source:* Penny Wieser

⊙ QUANTIFYING RESERVOIRS AND FLUXES

Five years after Berner's insightful overview of the deep carbon cycle, the Deep Carbon Observatory (DCO) set out to improve knowledge of the deep fluxes of carbon-bearing fluids to and from the mantle throughout Earth's history. An important driver of this mission was to improve our understanding of the chemical environments of carbon in extreme conditions, such as the solubilities of carbon in mantle silicates and the viscosity of carbon-bearing fluids.

Our historical account now takes us to the University of California, Davis, where the research of Louise Kellogg (1959–2019), who was professor of geophysics and director of computational geodynamics, focused on understanding the dynamics of the solid Earth. She and her colleagues developed advanced computational models that are firmly anchored in observation and informed by large, complex data sets. The initial conditions for their modeling include the present-day masses of carbon in the reservoirs, with data sourced from numerous investigators.[38] The core is by far our planet's largest carbon reservoir, packing a massive 4×10^9 Gt (1 Gt = 10^{12} kg). It has long been accepted that the inner core is crystalline iron alloyed with a small amount of nickel and diluted by some lighter elements, because the core's density is somewhat lower than the density of pure iron at high pressures and temperatures. The next largest reservoir is the mantle, at 2×10^8 Gt, which is 5 percent of the carbon in the core. For the continental crust, the Davis scientists used 4.2×10^7 Gt for carbon sequestered in carbonates. The oceans currently hold 4×10^4 Gt and the atmosphere 850 Gt, from which it is apparent that the oceans contain 98 percent of the carbon above the surface, with the atmosphere retaining just 2 percent. Organic carbon, mainly in the form of sediments and the biosphere, amounts to 10^4 Gt. The carbon in the core does not seem to play a significant role in the deep carbon cycle. For the atmosphere, the early Earth value (1.57×10^7 Gt) is derived by analogy to Venus; all but 850 Gt has been transferred to the mantle by subduction.

Kellogg's team considered the directions of flow between four large reservoirs: mantle, crust, ocean and atmosphere. Deep carbon participates in two principal transfers: it goes down by subduction and comes up through arc volcanoes and diffuse venting. Ébelmen, Högbom, Urey, Craig, Berner and Kellogg successively built on the work of predecessors to enable us to see that the mass of carbon involved in the familiar shallow carbon cycle reflects the balance between the up and down transfers of the deep cycle. Craig E. Manning of the University of California, Los Angeles has expressed this poetically:

> In essence, the carbon that is so essential to life and modern society is available to us only with the permission of the deep Earth.[39]

In 2015, Manning and Peter Kelemen of the Lamont–Doherty Earth Observatory gave an evaluation of carbon fluxes in the subduction zone. Their estimates indicate that almost all of the carbon in subducting plates and their oceanic sediments gets extracted in the melts and fluids. This carbon then rises in the form of buoyant ductile material or is transported in fluids and melts into the overlying plate. The study carried out by Kelemen and Manning suggested that this would mean that substantial quantities of carbon have been stored in the mantle lithosphere and crust rather than descending deep into the convecting mantle.[40] This raised the intriguing possibility that the continental lithosphere is a vast store for deep carbon.

A couple of years later, Stephen F. Foley at Macquarie University, New South Wales, and Tobias P. Fischer at the University of New Mexico, Albuquerque, modeled this storage capacity. Their analysis of the extent to which carbon in the lithosphere has increased and been mobilized by cycles of freezing and melting throughout geological history led them to conclude that Earth's deep carbon budget had been seriously underestimated. In their model, carbon-rich melts from the convecting mantle rise until they are blocked by a carapace of cold lithosphere, where they freeze out as carbon-bearing rock; later, mantle plumes turn up the heat and melt the rock; the volatiles are thus released, and carbon dioxide escapes through continental rifts. The two geochemists suggested that during the breakup of a supercontinent degassing through continental rift zones could result in the sudden release of CO_2 at a much higher rate than the rate of release from basaltic volcanism in the oceanic environment. This would lead to a short-term global warming incident triggered by tectonics.

⊙ 2009–2019: DECADE OF DEEP CARBON DISCOVERY

To conclude this chapter, I assess the geophysical and geochemical significance of the DCO's decade of discovery. The introduction of plate tectonic theory in the late 1960s ushered in the development of quantitative modeling of the geochemistry and geophysics of Earth's interior across geological time. It quickly became apparent that the archive of data on the transfer of carbon between reservoirs was grossly inadequate, and likely would remain so for decades.

Theorists making models are confronted with baffling complexity: multiple pathways feeding carbon in its many guises – gases, fluids, carbonates, etc. – between the mantle, the lithosphere, the crust and the atmosphere. As if that were not enough, bear in mind that the apparent elegance of the scheme conceals challenges of a kind that cannot readily be depicted in a two-dimensional diagram on a flat page. Temperatures and pressures increase dramatically as the sub-ducted slab descends, unleashing all kinds of chemical reactions and metamorphoses. Carbon-bearing matter in the interior exists in three phases – as gas, liquid or solid – so the chemical evolution of the subducting slab is determined by a multitude of phase changes that we'll meet in the next chapter.

In Chapter 6, we saw how the time dimension is another chal-lenge for modeling Earth's history. Modeling of carbon cycling at the surface is mostly considered in time steps of a year, a decade, a century or a millennium; now contrast that with carbon in the deep, where one thinks of millions or billions of years. Estimates of the *net* flux into or out of a reservoir are compromised by poorly constrained data. In part, that's because the pathways are not two-way streets in which the ins and the outs can be determined arithmetically. For example, Kelemen and Manning's survey of published data on the carbon being subducted estimated that the total input to subduction zones is 40–66 Mt per year and that the total output from those subduction zones is 14–66 Mt per year. The Fifth Assessment Report of the IPCC determined the total geological CO_2 flux at 370 Mt per year. In the same year, another review initiated by the DCO estimated that the global emissions were 50 percent higher, at 540 Mt per year.

I've selected these data for the purpose of illustrating the mag-nitude of the uncertainties in our knowledge of deep carbon and its fluxes. In 2011, the DCO launched its Deep Earth Carbon Degassing initiative – DECADE – in response to the urgent need to determine accurate global fluxes of volcanic CO_2 to the atmosphere. This strat-egy brought together both new and established technologies by setting up networks on 20 highly active degassing volcanoes. To complement this field network, the global collaboration undertook laboratory analysis such as elemental and isotopic mass spectrometry of volcanic gases. Automated laser sensors were deployed on crater rims to sample volcanic plumes, delivering the data by telemetry. The DECADE group encouraged formal collaborations among national volcano observatories on seven continents. These activities addressed "the most vexing problem of understanding the deep carbon cycle" through the coordinated efforts of interdisciplinary teams. At the time of writing, this research activity continues to be pursued with vigor.

REFERENCES

1. Le Pichon, X. Sea-floor spreading and continental drift. *Journal of Geophysical Research* **73**, 3661–3697 (1968).
2. Isacks, B. L., Oliver, J. and Sykes, L. R. Seismology and the new global tectonics. *Journal of Geophysical Research* **73**, 5855–5899 (1968).
3. Coltice, N., Gérault, M. and Ulvrová, M. A mantle convection perspective on global tectonics. *Earth Science Reviews* **165**, 120–150 (2017).
4. Perry, M. *Narrative of the Expedition of an American Squadron to the China Seas and Japan 1852–1854.* F. L. Hawks (ed.) (Congress of the United States, 1856).
5. Dickinson, W. R. The coming of plate tectonics to the Pacific Rim. An essay in Oreskes, Naomi Plate Tectonics: An Insider's History of the Modern Theory of the Earth (Westview Press, 2011), pp. 264–287.
6. Galvez, M. E. and Gaillardet, J. Historical constraints on the origins of the carbon cycle concept. *Comptes Rendus Geoscience* **344**, 549–567 (2012).
7. Encyclopedia.com. Spallanzani, Lazzaro. *Complete Dictionary of Scientific Biography.* Available at: www.encyclopedia.com/people/science-and-technology/biology-biographies/lazzaro-spallanzani (accessed April 22, 2019).
8. Spallanzani, L. *Voyages dans les Deux Siciles et dans quelque parties des Appenines.* **3** (H. Haller, 1796).
9. Spallanzani, L. Observations faites dans l'ile de Cythère en 1785. *Journal de Physique, de Chimie, d'Histoire Naturelle et des Arts* **47**, 278–283 (1798).
10. Wedgwood, J. XIX. An attempt to make a thermometer for measuring the higher degrees of heat, from a red heat up to the strongest that vessels made of clay can support. *Philosophical Transactions of the Royal Society of London* **72**, 305–326 (1782).

11. Davy, H. An essay on the generation of phosoxygen and on the causes of the colours of organic beings. In *Contributions to Physical and Medical Knowledge, Principally from the West of England*. T. N. Longman (ed.) (Thomas Beddoes, 1799), pp. 95–120.

12. De Saussure, N. T. *Recherches chimiques sur la vegetation* (Vanyon, 1804).

13. Boussingault, J. B. Recherches chimiques sur la nature des fluids élastiques qui se dégagent des volcans de l'équateur. *Annales de chimie et de physique* **52**, 181–189 (1832).

14. Boussingault, J. B. Recherches sur la composition de l'atmosphere. *Annales de chimie et de physique* **57**, 148–182 (1834).

15. Bischof, G. *Lehrbuch der Chemischen und Physikalischen Geologie* (Adolph Marcus, 1847).

16. Ébelmen, J. J. Sur les produits de la décomposition des espèces minerals de la famille de silicates. *Annales des Mines* **7**, 3–66 (1845).

17. Ébelmen, J. J. Sur la décomposition des roches. *Annales des Mines* **12**, 627–654 (1847).

18. Ébelmen, J. J. Recherches sur la décomposition des roches. *Comptes Rendus de l'Académie des Sciences* **26**, 38–41 (1848).

19. Hunt, T. S. On the Earth's climate in Paleozoic times. *American Journal of Science and Arts* **36**, 396–398 (1863).

20. Hunt, T. S. The chemical and geological relations of the atmosphere. *American Journal of Science* **19**, 349–363 (1880).

21. Högbom, A. G. On the probability of secular variations in atmospheric carbon dioxide. *Svensk Kimisk Tidskrift* **6**, 169–176 (1894).

22. Arrhenius, S. On the influence of carbonic acid in the air upon the temperature of the ground. *Philosophical Magazine and Journal of Science* **41**, 237–276 (1896).

23. Chamberlin, T. C. An attempt to frame a working hypothesis of the cause of glacial periods on an atmospheric basis. *Journal of Geology* **7**, 545–584 (1899).

24. Berner, R. A. and Högbom, A. G. Högbom and the development of the concept of the geochemical carbon cycle. *American Journal of Science* **295**, 491–495 (1995).

25. Harmon, C. The geochemistry of the stable carbon isotopes. *Geochimica et Cosmochimica Acta* **3**, 53–92 (1953).

26. Urey, H. C. On the chemical history of the earth and the origin of life. *Proceedings of the National Academy of Science* **38**, 351–363 (1952).

27. Berner, R. A. and Maasch, K. A. Chemical weathering and controls on atmospheric O_2 and CO_2. *Geochimica et Cosmochimica Acta* **60**, 1633–1637 (1996).

28. Berner, R. A. Jacques-Joseph Ébelmen, the founder of earth system science. *Comptes Rendus Geoscience* **344**, 544–548 (2012).

29. Libby, W. F. Atmospheric helium three and radiocarbon from cosmic radiation. *Physical Review* **69**, 671 (1946).

30. Anderson, E. C. *et al.* Natural radiocarbon from cosmic radiation. *Physical Review* **72**, 931 (1947).

31. Harmon, C. The natural distribution of radiocarbon and the exchange time of carbon dioxide between atmosphere and sea. *Tellus* **9**, 1–17 (1957).

32. Revelle, R. and Suess, H. E. Carbon dioxide exchange between atmosphere and ocean and the question of an increase of atmospheric CO_2 during the past decades. *Tellus* **9**, 18–27 (1957).

33. Berner, R. A., Lasaga, A. C. and Garrels, R. M. The carbonate–silicate geochemical cycle and its effect on atmospheric carbon dioxide over the past 100 million years. *American Journal of Science* **283**, 641–685 (1983).

34. Berner, R. A. and Lasaga, A. C. Modelling the geochemical carbon cycle. *Scientific American* **260**, 74–81 (1989).

35. Berner, R. A. *The Phanerozoic Carbon Cycle* (Oxford University Press, 2004).

36. Berner, R. A. Atmospheric carbon dioxide levels over Phanerozoic time. *Science* **249**, 1382–1386 (1990).

37. Berner, R. A. The long-term carbon cycle, fossil fuels and atmospheric composition. *Nature* **426**, 323–326 (2003).

38. Kellogg, L. H., Turcotte, D. L. and Lokavarapu, H. V. A box model for the transport of carbon between major carbon reservoirs over geologic time. In *American Geophysical Union, Fall Meeting 2018 Abstracts* (American Geophysical Union, 2018), Abstract DI33B-0034.

39. Manning, C. E. A piece of the deep carbon puzzle. *Nature Geoscience* **7**, 333–334 (2014).

40. Kelemen, P. B. and Manning, C. E. Reevaluating carbon fluxes in subduction zones, what goes down, mostly comes up. *Proceedings of the National Academy of Science* **112**, E3997–E4006 (2015).

13   Carbon-Bearing Phases in the Mantle

⊙ INVENTING HIGH-PRESSURE PHYSICS AND
   CHEMISTRY

Earth is the only known *habitable* planet in the solar system. Although there are half a dozen planetary moons that may have habitable zones beneath their surfaces, astronomers have yet to find an exoplanet with conditions deemed suitable for life on its surface. From a geological point of view, Earth's distinctive features include the presence of life, the abundance of liquid water, the long-term tectonic system and the profusion of organic carbon in contact with the oxygen-bearing atmosphere. The carbon cycle inextricably binds biological life at the surface with carbon on the move in the interior. Through laboratory experiments, we have discovered that, following subduction, carbon-bearing materials undergo great transformation in the high-pressure hothouse of the mantle. Earth's subduction factory plays a key role in the deep carbon cycle by feeding the mantle with different carbon phases, four-fifths of which are carbonates and one-fifth organic carbon, proportions that have remained relatively stable since Earth's biosphere became established.

The investigation of how carbon minerals behave in Earth's interior takes place in our physics and chemistry laboratories, where the high-pressure and high-temperature conditions from crust to core are reproduced. The history of this type of high-pressure physics begins in Oxford in 1659, when Irish physicist Robert Boyle (1627–1691) and his talented assistant Robert Hooke perfected a "pneumatical engine" for conducting experiments on the nature of air. Hooke's breakthrough led Boyle to propose his eponymous law on the pressure–volume correlation: PV = constant. Half a century later,

FIGURE 13.1 Percy Williams Bridgman, pioneer of high-pressure materials science. *Source:* Nobel Foundation Archives (1946)

in 1712, Thomas Newcomen (1644–1729) built the first practical steam engine for pumping water from mines. Britain's Industrial Revolution took off by harnessing high-energy steam power from carbon combustion. The first half of the nineteenth century witnessed the rise of classical thermodynamics, followed by the formulation of chemical thermodynamics in the 1870s thanks to the American mathematical physicist Josiah Willard Gibbs (1839–1903) of Yale College. The development of these two aspects of thermodynamics became essential to improving our knowledge of the geochemistry of the interior.

Percy Williams Bridgman (1882–1961) initiated the study of solid-state physics at high pressure (Figure 13.1). He enrolled at Harvard College in 1900, where, following his graduation, he spent 54 years of his scientific life in the department of physics. In 1946, he was awarded the Nobel Prize in Physics "for the invention of an apparatus to produce extremely high pressures, and for the discoveries he made therewith in the field of high-pressure physics." Bridgman strode along a path to fame with an unflagging passion for science, a penetrating and single-minded commitment to analytical thinking and exceptional ability in the mechanical arts. Being skilled at carpentry and plumbing, he effortlessly solved the ordinary problems of "household maintenance" in the lab. He wrote nearly 200 papers, mostly in a refreshing and direct first-person style. Bridgman never put together a large research group, although he did supervise graduate students. He dodged the research bandwagons and declined the dull duty of serving on university committees. We don't know what stimulated him to work on high-pressure physics because he left no record of his motivation. It must surely be relevant that Theodore Richards (1866–1928) in Harvard's chemistry department had measured the compressibility of elements at pressures of about 500 atmospheres (i.e. ~50 MPa), and Bridgman would have known about that.

Bridgman set out on his high-pressure adventures by peering at the effects of pressure on the refractive indices of liquids. To attain stability at higher hydrostatic pressures ($\leq$650 MPa), he invented a deformable pressure seal (or gasket) that was self-sealing: the greater the pressure he applied, the tighter the fit. He modestly noted this huge breakthrough in high-pressure instrumentation with a brief mention in a paper on pressure measurement.[1] To unlock the potential of this new leak-proof packing, he introduced a hydraulic ram in 1910, which pushed the practical limit for sustained pressure to 2 GPa. Thereafter, Bridgman's papers on the compression, electrical resistance and mechanical strength of materials poured out of the physics department. As soon as he had perfected this novel technique, he applied it to whatever he could lay his hands on and continued to ratchet up the pressure. He publicized his data in papers with catchall titles such as: "Rough compression of 177 substances to 40,000 kg/cm^2" (i.e. 4 GPa).[2] Bridgman's major contribution to deep carbon science essentially sprang from his confidence and singular ability to build and run dangerous, high-pressure experiments at lab scale without blasting the pressure chamber to smithereens. Bridgman routinely worked at pressures up to 10 GPa, achieving a record high of 40 GPa.

⊙ DIAMOND SYNTHESIS: SEEKING THE HOLY GRAIL

Diamond synthesis became the primary research driver of high-pressure research in the first half of the twentieth century. This commercial topic earns its place in this history of deep carbon science because businesses provided the funding for exploring petrology and mineralogy under extreme physical conditions. Even Bridgman became sidetracked by the possibility of making diamond. In 1941, he first trialed diamond synthesis with a 1000-ton commercial press, crushing graphite with pressures of up to 4.5 GPa and temperatures of 2000°C. No diamonds crystallized, but his data did sketch the phase diagram governing the transition of graphite to diamond, as well as the reverse transition of diamond to graphite. We'll return to this when we consider the stability of diamond in the mantle in the next chapter. After World War II, Bridgman accepted that his high-pressure skills did not extend to assembling the high-temperature technology necessary for diamond synthesis.[3] However, one of his commercial sponsors, the Norton Company of Worcester, Massachusetts, did continue an expensive all-out attack aimed at making commercial diamonds for use in high-performance abrasives. They relocated their 1000-ton press to their Worcester lab, where their employees engaged in world-class research into super-hard materials.[4] Bridgman continued to take an interest in their results, although he had to fend off many approaches from amateur scientists who sought his aid to turn their impractical ideas into untold riches.[5]

Loring Coes Jr. (1915–1978), an outstanding analytical chemist at Norton, and he is recognized today for his stunning success with the synthesis of minerals at high pressure. Bridgman had already shown that you must heat while you squeeze. Because heat fatally weakens steel, Coes eschewed the weapons-grade approach of adding a colossal steel press to his laboratory setup. He preferred to use small and perfectly crafted piston machines fabricated from tungsten carbide and an extremely hard ceramic form of aluminum oxide.

Norton's busy technicians supplied plenty of spares to Coes for making good when frequent breakages of his pressure cylinder occurred. Coes carried out a brilliant innovation by adding an internal electrical furnace made of graphite to the pressure vessel. With that clever improvement, the Norton team could now heat samples to more than 1000°C. They started cooking a wide range of mineral phases that only occur deep within Earth. This was a significant advance on Bridgman's more academic approach to research, which had largely focused on everyday materials. Because Coes worked for specific outcomes in an industrial laboratory rather than pursuing blue-sky research in a university, his mineralogical discoveries from 1947 to 1953 remained the private corporate assets of Norton, his employer. They carefully shielded the data from the scrutiny of inquisitive competitors, so no highlight papers appeared in academic journals.

Norton's operation blossomed under the talented leadership of Coes. He burrowed into the metamorphosis of mud, silt and sediments under the conditions likely to exist in the mantle. Such experiments generally cooked up mineral forms already familiar to geochemists, but that all changed abruptly late in 1952 when Coes worked on silica (silicon dioxide). Silica in the form of quartz grains is the commonest mineral on beaches. Coes tapped two-tenths of a gram of finely powdered silica and reagents into a small iron capsule 12 mm long and 5 mm in diameter. After baking for 15 hours at a pressure of ~3.5 GPa and a temperature of ~750°C, Coes cracked open the capsule: this exposed 20–30 mg of crystals never seen before. This new form of silica was about 10 percent denser and 50 percent harder than the commonest variety of quartz. This unexpected discovery validated a long-standing suspicion of geochemists that dense minerals, hitherto unknown, lurk in the high-pressure furnace of Earth's deep interior. Quite exceptionally, Norton granted Coes permission to publish a note in *Science* on the new mineral form of silica; it would later be named coesite in his honor. His announcement displays his characteristic modesty and dignity:

> [A] new dense form of crystalline silica has been discovered. The new silica has not previously been described as the product of synthesis nor has it been discovered in nature as a rock constituent. The conditions needed for the formation of the new dense silica, together with its great stability, may provide a means by which the conditions attendant on the crystallization of some deep-seated rocks can be more closely estimated.[6]

Coes was an outsider without any track record in academic research. Nevertheless, the pioneering paper piqued the interest of academics with its tantalizing news that "A subsequent paper on the synthesis of several naturally occurring minerals will greatly amplify this information." That set the scene for the geological community to follow up on Coes' high-pressure techniques of laboratory investigations for the exploration of Earth's interior.

Meanwhile, during the six years that passed while Coes refined his techniques for squeezing all manner of materials, two commercial rivals had risen to the diamond challenge. They were the Swedish general electric company ASEA and the General Electric Company (GE) in the USA. At ASEA, Baltzar von Platen (1898–1984) initiated the quest in 1941. An eccentric genius, he supported himself with the bounteous royalties flowing from his invention of gas absorption refrigeration (commercialized by Electrolux). For ASEA, he carried out a series of dangerous experiments with exotic machinery designed by himself. These inventions produced no diamonds whatsoever, forcing him to sell his diamond-making designs to ASEA in 1952 and leave the field. ASEA doggedly continued the work, which looked like a bad idea, until one Erik G. Lundblad (1925–2004) headed up the quest. Bob Hazen's pen-portrait of Lundblad described him as "a big, bullish extrovert with an appetite for big meals, strong drink, [and] good cigars." Quite a swashbuckler, then, as well as the enthusiast who crystallized ASEA's success.[4] Crucially, Lundblad changed the strategy in 1953: he advised the ASEA team to try a mixture of iron carbide (FeC) and graphite in the sample assembly. Lundblad's idea was that carbon dissolves in liquid iron at high temperature and crystallizes out when cooled at high pressure. On February 16, 1953, the ASEA team baked the mixture at 8.3 GPa for 1 hour. Microscopic examination revealed 40 or so minute diamond crystals. Tiny synthetic diamonds were found in two further experiments that year, which might have called for celebrations. No way! ASEA enforced an absurd silence, either for commercial reasons or, more probably, because it would have been embarrassing for them to admit publicly that the multimillion-dollar mission to produce amazing gems had yielded only a feeble harvest of teeny-weeny rough diamonds.

ASEA's rivals at GE had three different devices, and their creators vied for time on their one 400-ton press. Naturally, there were conflicts of interest, as well as internal disagreements about who reached the goal first. On December 16, 1954, Howard Tracy Hall (1919–2008) became the first person to grow a synthetic diamond by a reproducible, verifiable and witnessed process using a press of his own design. Fifteen years later, Hall described how he had conducted a 38-minute experiment at 10 GPa (i.e. 100,000 atmospheres) and 1600°C, and with trembling hands he broke open the capsule:

> Indescribable emotion overcame me and I had to find a place to sit down! My eyes had caught the flashing light from dozens of triangular faces of octahedral crystals ... and I knew that diamonds had finally been made by man.[7]

Within a few years, diamond-making at GE became routine. By the early 1960s, their technology could grow perfect crystals at the rate of one carat per week, and gem-quality crystals were achieved in the 1970s. Today, the production of synthetic diamond is a huge industry with numerous applications in optical materials, semiconductor electronics and machine tools.

Bridgman, Coes, Lundblad and Hall were the pioneering quartet whose research in high-pressure, high-temperature physics impacted positively on the field of deep carbon science. Bridgman set up the instrumental foundations, perfected the experimental techniques and measured the high-pressure properties of almost any material to hand. By so doing, Bridgman replicated the environment of the upper mantle. Coes' innovation was a method of internally heating the samples in the furnace to reach mantle temperatures. With those advances, both the pressure and the temperature in the mantle could be simulated in the lab. Lundblad's touch of eccentric genius was to use iron carbide. Hall made substantial improvements to the anvil press and succeeded with carbon mixed with iron. Overall, the pure science payoff from the high-pressure research of Bridgman and the diamond makers was the elucidation of the physical conditions in the mantle necessary for the formation of deep diamonds, as well as their subsequent safe delivery to the surface.

⊙ CARNEGIE GEOPHYSICAL LABORATORY: HIGH-PRESSURE MINERALOGY

I shall now revert back half a century to outline the history of high-pressure mineralogy at the Carnegie Institution of Washington, DC, which had established its Geophysical Laboratory in 1905, coincidentally the same year that Bridgman had begun such research at Harvard. The foundational purpose of the Geophysical Laboratory was to investigate the processes that control the composition and structure of Earth (as were known at the time), including developing the underlying physics and chemistry, as well as inventing and building the expertise and tools required for the task. That mission diversified a century later to embrace the physics, chemistry and biology of Earth over the entire range of conditions that our planet has undergone since its formation. The Geophysical Laboratory concentrated on experimental studies in petrology and geochemistry. There was, and still is, a science-driven emphasis on constructing the phase diagrams of mineral systems with several components, from which conclusions could be drawn on the origins and properties of igneous rocks. Carnegie scientists developed novel techniques for studying the behavior of rocks at high temperatures and pressures. Although there was some overlap with the work of Bridgman, the Carnegie agenda became distinctive because of its focus on rocks rather than materials science in general. In 1947, no less a figure than Victor Goldschmidt credited Carnegie scientists with the discovery of much of what was then known about the physical chemistry of silicates, which form the principal component of the lithosphere.[8]

The dazzling science career of Carnegie's Hatten (Hat) S. Yoder Jr. (1921–2003) originated in World War II when the US Navy had an urgent need for meteorologists. He worked with a team of Russian and American meteorologists, hunkering down in Siberia, who were charged with forecasting the weather conditions before the anticipated Allied invasion of Japan. A decorated officer, Yoder rose to the rank of lieutenant commander and served a further 12 years in the Reserves.[9,10] Following the completion of his doctorate in petrology at the Massachusetts Institute of Technology in 1948, he headed south to Washington, DC, for a 55-year association with the Geophysical Laboratory. Basalt was his lifetime interest, and his passion made him the world expert on its origin.[11] Coes' short paper in *Science* about dense crystalline silica greatly intrigued Yoder, who invited its author to Washington, DC, to give a presentation of the Norton research on minerals.

Following this presentation, 10 skeptics led by Yoder were invited to a site meeting at the Norton plant hosted by Coes on December 4, 1953. Why skeptics? Simply because, up to that day, they believed the formation of deep Earth minerals must involve complex processes operating over an immense timescale that could not be reproduced in the laboratory. In other words, they suspected that the lab simulations that Coes had described at the Washington presentation were only valid for near-surface rock formation.[4] Once they experienced the elegant look and sophisticated feel of Coes' equipment at the Norton plant, they changed their minds. His specimen crystals seemed quite unbelievable at first glance: many were large enough to pick up with tweezers. In addition, the X-ray diffraction patterns confirmed what minerals had been made. During the 10-hour train journey back to Carnegie, Yoder decided that the

Geophysical Laboratory had to move with the times by acquiring the expertise to create conditions in the lab akin to those of the lower continental crust and the upper mantle.

On December 8, 1953, Yoder composed on behalf of the skeptics a fulsome note of thanks to Coes:

> We all greatly appreciate the fine tour of your laboratory and the opportunity to see some of the details of your apparatus. You probably do not realize what a tremendous contribution you have made to the field of mineral synthesis and to geology in general. The impetus you have given to our work is very great.

With that acceptance of the evident richness of the mineralogy and geochemistry in the deep Earth, there was a call on geoscientists to embrace the new enterprise of high-temperature, high-pressure research. Sadly, the pioneer Loring Coes did not earn the recognition he deserved during his lifetime. Hazen has grimly cataloged the tragic moments of Coes' final 16 years: he was passed over for promotion by Norton; he suffered from depression and alcoholism; he was arrested for drunk driving; his marriage failed and he subsequently lost most of his possessions; and he experienced chronic ill health followed by an early death at the age of 63 from lung cancer triggered by his heavy smoking.

Meanwhile, Yoder worked heroically to construct a high-pressure, internally heated and very dangerous "bomb" capable of replicating the pressure and temperature conditions (1 MPa, 1200°C) under which igneous rocks and their constituent minerals had formed. The resulting papers and reviews on the origin and petro-chemical evolution of igneous rocks and the regimes governing metamorphism became some of the most highly cited references in the geosciences.[12] With respect to the history of deep carbon science, Yoder's research established the scientific platform for studies on heat transfer during the crystallization of silicate melts. This cast a new light on heat transfer in the upper mantle. The majority of the carbon in the crust and upper mantle today is present as carbon dioxide in geological fluids, and silicate melts are the main agents for transport-ing carbon from the interior to the surface.

⊙ GEOSCIENTISTS ARE ATTRACTED TO HIGH-PRESSURE RESEARCH

Results from the high-pressure research pioneered at Harvard by Bridgman as well as in Washington, DC, by the Carnegie scientists soon attracted recognition from geoscientists. When Arthur Holmes, for example, reviewed research on the thermal history of Earth in 1933,[13] he took care to cite three sources: Bridgman's textbook on high-pressure physics[14]; important research carried out in Göttingen by Gustav H. J. A. Tammann (1861–1938) on the solid state of matter at enhanced pressure and temperature[15]; and an influential paper on discontinuities in the mantle by two outstanding seismologists at Caltech, Beno Gutenberg (1889–1960) and Charles Francis Richter (1900–1985).[16] High-pressure studies drew attention to the remark-able effect that water has on the strength of rock in Earth's interior. If water is incorporated into a crystal, it acts as an impurity that lowers the melting point; the water is not present as a liquid, but as individ-ual molecules infiltrating the mineral's atomic structure. What geo-scientists learned from the early period of high-pressure research was that the silicate mantle rock, composed of the minerals olivine, garnet and pyroxene, is not actually molten (as is often asserted in high school textbooks), but plastic. But more importantly, it became clear that water lowers the melting point (solidus) of peridotite from ~2000°C at 200 kilometres to 1300–1800°C.

As the results from high-pressure research gradually accumu-lated, it became abundantly clear that geochemical reactions in the mantle are dramatically different from those at the surface. In the twentieth century, two developments in scientific thought opened up the chemistry of the interior to interrogation: the introduction of extreme physics in laboratories for the replication of the mineralogy of the mantle and the plate tectonic revolution, which, together with the invention of the diamond anvil cell (Figure 13.2), led to a better

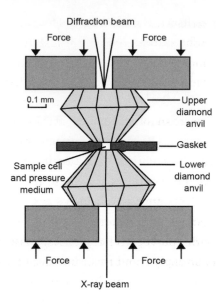

Diffraction beam

Force   Force

0.1 mm

Upper diamond anvil

Gasket

Sample cell and pressure medium

Lower diamond anvil

Force   Force

X-ray beam

FIGURE 13.2 Diamond anvil cell. Bridgman transformed high-pressure research with his development of an opposed anvil device in which two anvils made of tungsten carbide were pressed against one another with a lever arm. In 1959, a group at the National Bureau of Standards created the first diamond anvil cell by substituting diamond for tungsten carbide. The sample can be viewed through the diamond and heated by laser beam. Today (2020), the highest pressure attainable is 640 GPa and the highest temperature attainable is almost 6000 K: these conditions are more extreme than those at the center of Earth.
Source: Penny Wieser

understanding of the physical and chemical changes affecting our planet over a period of billions of years. As I have noted before, from the mid-twentieth century, education and research in the geosciences expanded enormously, along with space science, nuclear physics, molecular biology, information technology and so on. These trends in knowledge acquisition and application have made interdisciplinary, cooperative research indispensable. Consequently, the narrative history of the geosciences since the mid-twentieth century is necessarily focused on broader trends within the global geoscience community, rather than the pivotal achievements of brilliant individuals.

⊙ FIVE REACTIONS AND THE DEEP CARBON CYCLE

One theme tracked in this chapter is the discovery of the chemical reactions that play an important role in the deep carbon cycle. During 2015–2019, Jie Li of the University of Michigan and Simon Redfern at the University of Cambridge ran an imaginative project to pinpoint these key reactions.[17] They conducted their inquiry through the Deep Carbon Observatory's international network of more than 1000 scientists. An initial call for suggestions provoked 120 submissions, covering a wide range of passions and enthusiasms. Arising from the call, participants at a two-day workshop held in March 2018 at the Carnegie Institution for Science were charged with winnowing down the submissions to five reactions seen as being central to the deep carbon cycle and the evolution of life on Earth. In the setting of the federal capital, Washington, DC, it was surely appropriate that the 50 representatives of Deep Carbon Observatory communities and researchers debated openly, with "provocative responses," assertive statements such as "I feel very strongly about this," and disputations that "were passionate and sometimes intellectually divisive." Invoking an anonymous voting procedure, the colleagues "converged on a set of reactions central to defining Earth." Yes, they did it by voting, not peer review or citation analysis. That's how decision-making is done in the District. The five reactions – or "winners" – topping the poll were: "hydrogenation, carboxylation, carbonation, carbon dioxide dissolution, and hydration." All quite a mouthful to recite, but I'll attempt give a rendering that's easy to follow in terms of the history of science. Taken together, these reactions provide a handy framework for understanding the complexity of the deep carbon cycle. As a brief example of the importance of the reactions I have chosen serpentinization, which uses three of these reactions, as we shall now see.

Serpentinization is mainly associated with ultramafic rocks, and it is central to our understanding of both the deep carbon and the deep water cycles. By definition, an ultramafic rock has a low silica content (less than 45 percent) and high magnesium and iron contents. The mantle is primarily composed of ultramafic rocks, with the ferromagnesian silicates olivine and pyroxene being dominant. Serpentinization is an umbrella term for an important set of geochemical reactions that result in the transformation of olivine and pyroxene over a wide range of pressure, temperature and catalysis conditions. This process – hydrothermal alteration of mantle rocks by hydration – occurs when ultramafic rock meets seawater and dissolved carbon dioxide at temperatures below ~400°C. This can happen, for example, when seawater invades the ocean crust through cracks and faults. The key chemical reagent here is iron (Fe^{2+}): it reduces water to produce hydrogen (hydrogenation) and hydrocarbons, as well as magnetite (Fe_3O_4) and a hydrated magnesium–iron silicate called serpentine, after which the process is named. Visually, serpentine is dark green, sometimes mottled and resembles the skin of a snake (serpent), hence the name. The serpentinization of ultramafic rock occurs in the mid-ocean ridges and forearc systems. It's a geochemical process that results in the injection of a variety of gas and fluid species into the interior: serpentinization reduces carbonates (carbonation reaction in reverse) to produce hydrogen, methane and a range of organic compounds.

The process, along with its intertwined reactions, is so multi-plexed that one can easily become baffled when trying to explain it using plain English. In simple terms, serpentinization takes place at relatively low temperatures for a metamorphic process, and the reaction is highly exothermic, pushing up rock temperatures by as much as ~250°C, which may lead to partial melting. Arthur Holmes, writing in 1927, 1928 and 1933, acknowledged that continental drift models required partial melting in the asthenosphere. He looked to radiogenic heating, but the view today is that serpentinization is the source of thermal energy. The mineral alteration is particularly significant at tectonic plate boundaries, where seawater seeps to depths far too deep to be accessed by current technology. Regardless of its complications, serpentinization has been widely accepted as the major pathway for producing molecular hydrogen (H_2) and methane (CH_4) over a huge range of environmental conditions below the crust and in the mantle. The molecular hydrogen, once released, has the potential to react with carbon species, leading to the assembly of organic molecules. It is abundantly clear that serpentinization significantly influences the reservoirs and fluxes of deep carbon at subduction zones and mid-ocean ridges.

The completion of the plate tectonic revolution in the late 1960s led to a prolific amount of research and discussion on serpentinization, and that torrent of papers continues unabated. The main points of discussion over the past half-century have been: the temperature–pressure regime of hydration; the mineral assemblages produced during serpentinization; the influence of fluid compositions on the mineralogy of serpentinite; and the origin of the serpentinizing fluids.[18]

⊙ CARBON MINERAL EVOLUTION

Chapters 2 and 3 showcased the long history of carbon's origin in stars and its subsequent transport to and incorporation within early Earth. We've seen how deep carbon exists in the atmosphere, lithosphere and mantle in many forms and phases. One story that runs through this history of deep carbon science is the role of evolution, and particularly the impact of the growth in complexity over long periods of time. In the geosciences, William Smith, Lavoisier, Georges Cuvier, Charles Lyell, James Hutton, Adam Sedgwick, Holmes and many others came to understand that the past is indeed foreign: life evolves, societies move on, the scientific method of inquiry adapts to new protocols, complexity and diversity displace simplicity and conformity, and so on. Chapter 7 introduced the history of radiometric dating and how the discovery of radiogenic heating confirmed the story of Earth's great antiquity.

In 1910, Robert Strutt (1875–1947) of Imperial College London wrote up the results of an exceptionally demanding experiment he had carried out to measure the residual helium resulting from radioactivity in samples of thorianite (mainly ThO_2, but with traces of other metallic oxides) and pitchblende (uraninite, UO_2). He offered a "Minimum Estimate of their Antiquity" of 280 million years for the specimens of thorianite, for which he had previously guesstimated 240 million years by an indirect and inexact method. One historical consequence of Strutt's confident measurement of the extreme antiquity of a specific *mineral* was the recognition that Earth's crustal mineralogy must surely have evolved over geological time.[19] Norman Levi Bowen (1887–1956) latched onto that theme of mineral evolution. He became a great pioneer of experimental petrology through his research on the evolutionary sequence of silicate minerals in igneous rocks. Much of his life's journey in science was at the Carnegie Geophysical Laboratory, where, in 1915, he became an international figure (at the young age of 28) thanks to his experimental findings on melting phenomena in silicate rocks. The results of Strutt and Bowen implied that the crustal mineralogy of Earth has evolved enormously over the past 4.5 billion years. This evolution has caused great changes in the surface composition and distribution of minerals.

Robert Hazen concisely expressed a strong interest in the "mineral kingdom's evolutionary narrative" when he wrote that, more than any other element, "carbon exemplifies the processes of mineral evolution." According to Hazen, four episodes have contributed to the fascinating historiography of carbon-bearing materials. The first mineral (crystal) in the universe was probably diamond, the subject of the next chapter. The second episode was formation of carbon-bearing stuff in the solar nebula, as preserved in meteorites, the topic already explored in Chapter 3. The third episode of carbon mineral evolution Hazen identified is carbon's participation in the richly dynamic geophysics and geochemistry of the crust and mantle, with its range of intensive variables such as temperature and pressure. And finally, the most recent addition to the narrative of this deep science has been carbon's profound influence on near-surface mineralogy through the evolution of the biosphere: living systems generate far from equilibrium conditions.[20,21]

Hazen pointed out in 2015 that the concept of mineral evolution leads to further considerations that enlarge our knowledge of Earth's history. The number of mineral species that exist on Earth has increased monotonically as a result of physical, chemical and biological processes. Hazen's consortium estimated that Earth might have hosted 420 mineral species in the Hadean Eon (4.6–4.0 Ga), with a further 1500 species added in the next billion years (Archean Eon, 4.0–3.0 Ga), during the course of which continents formed and the earliest microfossils were deposited (~3.5 Ga). By 2019, the International Mineralogical Association's list included ~5400 valid mineral species, with additions accumulating at a rate of ~50 annually. Hazen and his colleagues engaged in a debate on the philosophical aspects of mineral diversity, distribution and biological processes. Their review focused on "mineral ecology," a philosophical concept they introduced as a way of considering aspects of chance versus necessity (or contingency versus determinism) in the development of mineral diversity.[22] Although carbon is universally cited as being *necessary* for any plausible biochemistry, molecular biologists have yet to settle whether details of the genetic code are contingent (depending on chance) or deterministic (inevitable). When considering mineral evolution, Edward Grew and Hazen noted that chance and necessity must each play an important role in increasing mineral diversity.[23] They made a case study of beryllium minerals and concluded that on any Earth-like planet the abundant beryllium minerals would *necessarily* be associated with aluminum and silicon (e.g. beryl, $Be_3Al_2Si_6O_{18}$), whereas the rare minerals would result from chance association with rare elements, coupled with contingent events such as fortuitous preservation. They viewed these latter as "frozen accidents," and speculated that "dozens of rare beryllium minerals await to be discovered."

⊙ THE CARBON MINERAL CHALLENGE

Hazen's creative concept of mineral evolution and ecology paid off handsomely when he and his team moved on from beryllium to the carbon-bearing minerals. By employing data-mining along with well-established statistical filtering protocols, they invented a modeling approach to predict the total number of mineral species on Earth today: *minerals known* and *minerals missing*.[24] I'll adapt Hazen's description of their initiative in solving the quandary of how many minerals exist. Let's start with the obvious: minerals are solid-state assemblages of chemical elements. Those chemical elements vary widely in abundance and affinity to associate with other elements. Therefore, a correlation should exist between the crustal abundance of an element and how many species incorporate that element. As evidence for this, silicon, sulfur and aluminum each occur in more than 1000 *known* species. But, in a nutshell, let's now ask: How many minerals remain to be discovered? Hazen and colleagues used a light touch to crack that hard nut by searching for links between mineral diversity and distribution. They quizzed their quantitative data by inquiring, "Where we find this, can we find that?" It became a quest for *known unknowns* among the mineral species: they were classified as *known* from the use of analysis that predicts the outcome of rare events, but they were *unknown* because they were in the category of "undiscovered in any locality." This forensic procedure was analogous to that used by literary scholars striving to detect the period and authorship of an anonymous manuscript by sifting for rare patterns in the words and phrases. On "missing" minerals, the group predicted that many hundreds of minerals remained to be discovered, including 145 previously undescribed carbon minerals. Hazen and his associates analyzed the diversity distribution of 403 carbon-bearing minerals already known. This research led them to the unknowns, and it also flagged up suspected locales, such as quarries and caves where the suspects might be hanging out with other rare carbon minerals already on their database.[25]

This focus on undiscovered carbon minerals that have been crystallized in the deep was driven by Hazen's pioneering concept of mineral evolution, which narrates how life and geology have been intertwined throughout Earth's billions of years. Another impetus came from outer space. Planetary systems are everywhere we look in the Milky Way. It is teeming with them. But are they teeming with life? Astrobiologists crave Earth-like planets, but what does *Earth-like* actually mean? That question is quite subtle. Yes, an *Earth-like* planet has to orbit the "right kind" of star in its "habitable zone" and so on.

There's no shortage of them: in November 2013, astronomers reported that there could be as many as 10 billion of these in the Milky Way. But would any of them have evidence of actual life? When discussing the possibility of the detection of alien life, Carl Sagan had poetically proposed the aphorism that "extraordinary claims require extraordinary evidence." In so doing, he had unconsciously followed in the footsteps of Thomas Jefferson (1743–1826), third president of the USA and the principal author of the Declaration of Independence, who, on February 15, 1808, penned a letter to one Daniel Salmon, owner of a recently fallen meteorite, in which he averred that "verity needs proofs proportioned to their difficulty."[26] Today, our searches for planets with evidence of life therefore needs to take into account what we have learned about the dynamic deep carbon cycle that must be present for life to emerge. An inventory of an exoplanet's carbon-bearing minerals might indicate whether it ever had the potential for dynamic carbon and water cycles to initiate and sustain life.

Researchers at Carnegie launched their Carbon Mineral Challenge in 2015. This citizen science project mobilized amateur and professional collectors to find the missing minerals. The scientific goals incentivizing the quest were summarized in a series of questions that needed answers in order to recount the rich story of Earth and its carbon. What were the earliest carbon-bearing minerals on Earth? Do carbon-bearing minerals play a role in the origin of life? How did the evolution of life affect the evolution of Earth's carbon mineral inventory? Hazen and his colleagues suggested the hunters search around evaporated saline lakes, coal mine fires and dumps and volcanic fumaroles for the elusive minerals. They warned collectors of several challenges: a missing mineral might be as small as a grain of sand, or as ephemeral as a volcanic eruption. The predicted minerals included many hydrous carbonates that were expected to appear as small, colorless and relatively soft crystals in association with other carbonates: in short, rare and difficult to recognize.

The early discoveries included the following: abellaite, a sodium–lead carbonate from Spain; marklite, a hydrous copper carbonate from a German mine dump and identified from a single specimen; and four new carbon-bearing minerals containing uranium – two from the Czech Republic and two from Utah.[27] One headline-grabbing discovery was that of tinnunculite: this mineral gets fired up when excrement of the European kestrel (*Falco tinnunculus*) bakes in the hot gases from an underground coal fire. It was not specifically predicted by the Hazen collaboration, but it was in a location that had been fingered as a good place to look for undiscovered carbon minerals. When the Carbon Mineral Challenge finished in October 2019, it had documented 28 new carbon minerals in a variety of places: subarctic peaks, desert cliffs, a Chilean guano deposit and abandoned mines – all of which confirmed that the ecology of carbon minerals does indeed evolve with time, as our planet weathers the landscape, subducts the seafloor, recycles volatiles through the mantle and returns them to the surface through volcanic activity. One take-home lesson for the exobiologists is that the origin of life seemingly does require extraordinary geochemical, geophysical and geobiological circumstances in the deep carbon environments. According to Carnegie scientists, "The heart of habitability lies in the planetary interior."[28]

REFERENCES

1. Bridgman, P. W. The measurement of high hydrostatic pressure. II. A secondary mercury resistance gauge. *Proceedings of the American Academy of Arts and Sciences* **44**, 221–251 (1909).

2. Bridgman, P. W. Rough compression of 177 substances to 40,000 kg/cm². *Proceedings of the American Academy of Arts and Sciences* **76**, 71–87 (1948).

3. Bridgman, P. W. An experimental contribution to the problem of diamond synthesis. *The Journal of Chemical Physics* **15**, 92–98 (1947).

4. Hazen, R. M. *The Diamond Makers* (Cambridge University Press, 1999).

5. Bridgman, P. W. Synthetic diamonds. *Scientific American* **193**, 42–47 (1955).

6. Coes, L. A new dense crystalline silica. *Science* **118**, 131–132 (1953).

7. Hall, H. T. Personal experiences in high pressure. *The Chemist* **July**, 276–279 (1970).

8. Goldschmidt, V. M. *Geochemistry*. **78** (LWW, 1954).

9. Yoder, H. S. *Planned Invasion of Japan, 1945: The Siberian Weather Advantage.* **223** (American Philosophical Society, 1997).

10. Ernst, W. G., Hazen, R. M. and Mysen, B. *Biographical Memoir of S. Yoder Jr. (1921–2003)* (National Academy of Sciences, 2014).

11. Yoder, H. S. *Generation of Basaltic Magma* (National Academies, 1976).

12. Yoder Jr, H. S. and Tilley, C. E. Origin of basalt magmas: an experimental study of natural and synthetic rock systems. *Journal of Petrology* **3**, 342–532 (1962).

13. Holmes, A. The thermal history of the earth. *Journal of the Washington Academy of Sciences* **23**, 169–195 (1933).

14. Bridgman, P. W. *The Physics of High Pressure* (G. Bell & Sons, 1931).

15. Tammann, G. *Aggregatzustände: Die Zustandsänderungen der Materie in Abhängigkeit von Druck und Temperatur* (Leopold Voss, 1923).

16. Gutenberg, B. and Richter, C. F. On supposed discontinuities in the mantle of the earth. *Bulletin of the Seismological Society of America* **21**, 216–223 (1931).

17. Li, J. *et al.* Deep carbon cycle through five reactions. *American Mineralogist* **104**, 465–467 (2019).

18. Moody, J. B. Serpentinization: a review. *Lithos* **9**, 125–138 (1976).

19. Strutt, R. J. Measurements of the rate at which helium is produced in thorianite and pitchblende, with a minimum estimate of their antiquity. *Proceedings of the Royal Society of London. Series A, Containing Papers of a Mathematical and Physical Character* **84**, 379–388 (1910).

20. Hazen, R. M. *et al.* Mineral evolution. *American Mineralogist* **93**, 1693–1720 (2008).

21. Hazen, R. M., Downs, R. T., Kah, L. and Sverjensky, D. Carbon mineral evolution. *Reviews in Mineralogy and Geochemistry* **75**, 79–107 (2013).

22. Hazen, R. M., Grew, E. S., Downs, R. T., Golden, J. and Hystad, G. Mineral ecology: chance and necessity in the mineral diversity of terrestrial planets. *The Canadian Mineralogist* **53**, 295–324 (2015).

23. Grew, E. S. and Hazen, R. M. Beryllium mineral evolution. *American Mineralogist* **99**, 999–1021 (2014).

24. Hazen, R. M. *et al.* Earth's "missing" minerals. *American Mineralogist* **100**, 2344–2347 (2015).

25. Hazen, R. M., Hummer, D. R., Hystad, G., Downs, R. T. and Golden, J. J. Carbon mineral ecology: predicting the undiscovered minerals of carbon. *American Mineralogist* **101**, 889–906 (2016).

26. Jefferson, T. Letter to Daniel Salmon (1808).

27. The Carbon Mineral Challenge. *Rock&Gem* (2016).

28. Shahar, A., Driscoll, P., Weinberger, A. and Cody, G. What makes a planet habitable? *Science* **364**, 434–435 (2019).

14 Diamond in the Mantle

⊙ DELVING DEEP FOR DIAMOND

Diamonds form deep in the mantle, where they mostly remain unless propelled to the surface in a volcanic eruption. Diamond is by far our most important mineral messenger for discovering the history and chemistry of Earth's convecting mantle. Diamond provides a window to otherwise inaccessible geological processes that churned away 100 kilometres underground and often occurred billions of years ago.[1] Chemically, diamond is exceptionally pure: 99.9 percent or more elemental carbon. Diamond lasts almost forever, being the hardest known natural material by quite a margin. And since 1948, De Beers has used the marketing tagline "A diamond is forever" continuously to promote its diamond engagement rings.

The prehistory of the diamond trade suggests that the first discoveries of these shiny grains were made between 2500 and 1700 BCE by the Indus Valley Civilization in the Bronze Age. A diamond industry emerged in India during the fourth century BCE, when Sanskrit texts chronicled their use as a store of value and a means of trade. These texts explained how to grade a diamond's quality and characteristics. In the ancient world, diamond was prized: in India as a jewel symbolizing the rank of a ruler or as the eyes in sacred statues; in the West as a conspicuous statement of lifestyle and wealth. It was also prized for its magical power, attributed to its indestructability, and, when forcefully smashed, for the utility of the shards for engraving iron and making sharp implements. In the King James Bible, Jeremiah 17:1 proclaims: "The sin of Judah is written with a pen of iron and the point of a diamond ..." For two millennia, diamonds, as well as all manner of luxury goods from India, were traded along the

Karakoram and Hindu Kush route of the Silk Road. It was not until the eleventh century that diamond jewelry graced the royal courts of Europe. In colonial Brazil in the early 1700s, diamonds were accidentally discovered in panned river gravels, which sparked a chaotic diamond rush to the alluvial deposits in the state of Minas Gerais. After Portugal regulated its colonial trade from the middle of the seventeenth century, stones flooded into Amsterdam and London for the diamond cutters. Minas Gerais then dominated the global trade for nearly 150 years, until the meteoric rise from the 1880s of South Africa as the world's major producer. A handful of ancient texts recount that diamonds existed in northern Africa. Pliny's *Naturalis Historia* mentions Ethiopia as a source of diamonds. In the collected letters of the natural philosopher and courtier Lorenzo Magalotti (1637–1712), it is recorded that a merchant from Constantine, the ancient city in eastern Algeria, brought to Tuscany a cut and engraved diamond half the size of a hazelnut.[2]

In southern Africa, shortly before Christmas 1866, Erasmus Jacobs (1851–1933), the teenage son of a poor Boer farmer, was playfully pocketing pretty pebbles scattered along the south bank of the Orange River, near the small settlement of Hopetown. In this hot and desolate location in the Northern Cape province, sticks and stones were the natural playthings for children. Erasmus became very excited by a white stone that sparkled in the Sun. His mother, intrigued by her boy's find, showed it to a neighboring farmer, Schalk van Niekerk (1828–1887), also a colonial settler. Being fairly sure it was diamond, he "took care of it." After passing through other hands, the shiny pebble eventually dropped onto the desk of one Lorenzo Boyes (1836–1918), the acting Civil Commissioner (an officer appointed by the Colonial Service with responsibility for local government and justice). He pronounced it a diamond, and for confirmation dispatched it 300 kilometres south to William Atherstone (1814–1898), the foremost geologist at the Cape Colony. Following several tests, including asking a neighbor to see if it could cut glass, Atherstone replied to Boyes:

> I congratulate you on the stone you have sent me. It is a veritable diamond, weighs 21 carats, and is worth $714. Where that came from there must be lots more.

This child's chance find was mailed to London, where it was dubbed *Eureka*. Queen Victoria's crown jewelers confirmed that it was a diamond, but they did not trust the story of its provenance because they were cagey about being swindled by unscrupulous colonial traders. Sir Roderick Murchison (1792–1871) likewise was wary of falling for a tall story, saying that although it was indeed a diamond, the exaggerated claim that southern Africa was on the brink of becoming a major source of the gems could not be taken seriously unless and until diamonds were to be found *in situ* in the matrix.

As it turned out, there were "lots more." Once the *Eureka* news broke, optimistic prospectors raced to the scene, shoveling river gravel into rocker boxes and sluicing it to wash out gemstones. But on finding only a few small crystals, they became dejected and drifted away. Soon, in 1869, everyone's prospects brightened dramatically when a local shepherd (a *Griqua*) picked up an 83.5-carat (16.7-g) beauty on the riverbank. Instantly recognizing that it was a diamond crystal, van Niekerk seized the chance by trading almost all of his possessions (500 sheep, 10 oxen and a horse!) for the find. He then passed it to a local jeweler for £11,200, who forwarded it to diamond jewelers in Hatton Garden, London, where it was cut into a fabulous gemstone of 47.69 carats (9.54 g): the *Star of South Africa*. Currently in private ownership, it also known as the Dudley Diamond, and should not be confused with Cullinan I, the *Great Star of Africa*.[3]

Further stunning finds were unearthed in May 1871 some 30 kilometres southeast of the Vaal, at a farm owned by two Boers: Diederik (1862–1928) and Nicolaas (1830–1883) de Beer (*sic*). These pioneering prospectors found plenty of easily extractable diamonds in the "dry diggings," where there was no accessible water; they wondered if these diamonds were alluvial deposits left by an ancient flood of the Vaal. Within a couple of months, the de Beers' farm was again in the spotlight: diamonds were gouged out of Colesberg Kopje, a low hill 3.5 kilometres to the west of the initial excavations. The de Beer farmland thus became the birthplace of two enormously success-ful diamond enterprises: De Beers Mine and Kimberley Mine. News of the second source impelled another dash for diamond, when eager diggers, dodgy chancers and the usual camp followers crowded the site of what would become the richest treasure house in the world. Because the de Beer brothers could not hold back the growing tide of intruders trashing their land, they relinquished it for £6300. At that moment, nobody had any idea of the incalculable wealth just tens of metres beneath their feet because they all thought they were grubbing up alluvial diamonds. The moniker "De Beers" stuck to the mine, and it survives as the eponymous company that controls about three-quarters of global diamond mining today.

The Kimberley diggers declared an independent republic in 1870 and kicked out the Transvaal officials who had been main-taining a vestige of law and order. When the republicans pleaded with the British imperialists for protection, the Colonial Office responded favorably. London had a vested interest in gaining full control of the minerals in the South African Republic. The British authorities bought some order by paying off and evicting the Afrikaners (Boers, white farmers of Dutch descent) and native Africans (contemptuously referred to as Kaffirs). By December 1870, there were 10,000 British settlers in the short-lived Diggers' Republic. Prospectors purchased one or two claims of about 10 m^2, inside which they jury-rigged contraptions of ropes, pulleys and barrels for hauling out the diamondiferous ore they had loosened by pick and shovel. As soon as the soft ore crashed onto the sorting tables in the blazing sun, it disintegrated to a dry crumble, into which the diggers delved for glistening diamonds. This sifting for crystalline carbon from the deep soon threw up a grim and chaotic landscape, with a good deal of rough-and-ready justice for cheats and thieves. In response, on October 27, 1871, Queen Victoria's Colonial Officers deposed the republic and imposed imperial rule.

They declared 40,000 km^2 as British sovereign territory, which later merged with Cape Colony.[4]

Victoria famously loved diamonds. That's perhaps why, when the British conquered the Sikh empire in 1849, the punitive peace treaty they imposed required the following: "The gem called the Koh-i-Noor ... shall be surrendered by the Maharajah of Lahore to the Queen of England."[5] Victoria's beloved husband Albert had this unattractive and irregular 186-carat stone magnificently recut as a 105-carat gem, set as a personal brooch he gifted to his wife: "[A] trophy of the glory and strength of Your Majesty's Empire in the East."[6] Today, you can see this gem in the Tower of London, set in Queen Mary's coronation crown.

⊙ DIAMOND IN KIMBERLITE

At this stage, an adventurous mineralogist enters our story: Emil Wilhelm Cohen (1842–1905), who had studied physics and chemistry in Heidelberg and Berlin from 1863. As an assistant at the Mineralogical Institute in Berlin, he made geological surveys along the Rhine. In 1872, a firm of gem-dealers in Hamburg engaged Cohen for an in-depth probe of the extraordinary claims about the diamond fields of South Africa. We can be fairly sure that when Cohen arrived at the diamond diggings, he was the first resident European visitor to do so. He noticed that the earliest workings on the low mound (kopje) were dug in highly weathered rock known as "yellow ground." Cohen discerned that the surface hollows had originally been filled with a decomposed tuff-like rock enclosing diamonds and xenoliths – fragments of entrained minerals. His assessment of the geomorphology of the dry diggings was that they were originally steep-sided cylindrical columns that had served as the volcanic vents for the escape of magma and gases. This led him to the remarkable conclusion that "pipes" from the interior had rocketed diamonds, xenoliths and tuff to the surface by eruptive volcanic action in the deep past.

Edward John Dunn (1844–1937) was the next geologist to tread the diamond fields. When he was aged four, his parents had migrated to the colony of New South Wales, enduring a three-month sea journey. From boyhood, he had eagerly collected minerals and rocks, so it's perhaps not surprising that by the age of 20 we find him training as a mining surveyor. Two years after qualifying, he departed Australia for Africa, where, from 1871, he served as the government geologist for Cape Colony. Right away, surveying the geological structure of the diamond fields became his top priority, although he also became deeply involved with investigating the goldfields of Transvaal and the coal deposits of Cape Colony. In December 1873, he read a paper to the Geological Society, London.[7] He began:

> The conditions under which diamonds occur in South Africa are quite different from those in every other known locality and are so unusual as to deserve the earnest attention of all geologists.

His conclusions concurred with Cohen's: that the dry diggings were perched atop pipes in the shale that had carried diamonds to the surface; "it is reasonable to infer that they are merely the channels that connected ancient volcanic craters with deep-seated reservoirs of molten rock." Dunn's detailed mineralogical inspection of the rocks led him to infer that at the greatest depth then reached by the mines, the rock was unaltered, showing its original texture. In the decomposed rock he noted that although there were beautiful crystals of up to 150 carats,

> ... a large percentage consists of fragments and broken crystals; and it is noticeable that the corresponding pieces, even when the original crystal was of large size, are never found, though most carefully looked for.

From this, he concluded that the diamond-bearing rock was not the original matrix or "mother rock." All the same, though, he felt "it is marvellous that it contains such a rich sprinkling of glittering gems." What was clear, however, was that diamonds were only ever being found in the strange rock of the dry diggings.

In the oral discussion that followed, Nevil Story-Maskelyne (1823–1911), Keeper of Minerals at the British Museum, congratulated

Dunn for "showing that rock in which the diamonds occur is confined to the pipes." The British Museum had recently received rock specimens from all four pipes, sampled from the surface to a depth of 50 metres. Museum staff had subjected them to careful mineralogical examination, from which Maskelyne concluded that the host rock had a unique makeup of minerals. There were many rock fragments present, and he thought the shattering of diamonds was "due to the breaking up of the rocks in which the diamonds are now found." From the minerals he found, Maskelyne concluded that the diamondiferous rock was "undoubtedly once igneous rock." Furthermore, the "distinctive character of the diamonds" suggested that the source of diamonds was "not remote" from where they were found.[8] In recognition of his meticulous and pioneering fieldwork in South Africa and Australia, the Geological Society awarded Dunn its Murchison Medal.[9]

In 1886, a dozen years after Dunn's presentation, American mineralogist and geologist Henry Carvill Lewis (1853–1888) shone more light on the mineralogical properties of the rock-like "blue ground," which was by then being described as a remarkable porphyritic peridotite. Porphyritic rocks form when a column of rising magma undergoes slow cooling deep in the crust, followed by a period of rapid cooling at shallow depths. Lewis enthused about the Kimberley and De Beers mines, where the "blue ground has now been penetrated to a depth of 600 feet and is found to become harder and more rock-like as the depth increases." Lewis identified the diamond-bearing pipes as "true volcanic necks, composed of very basic lava associated with a volcanic breccia ... and that the diamonds are secondary minerals."[10] A year later, he gave a short paper on the geochemistry and mineralogy of the "matrix of diamond." Sensing that diamond-bearing rock deserved a distinctive name, he proposed kimberlite, after Kimberley. Lewis perceived a clear link between alluvial and kimberlite diamonds: in Borneo, New South Wales, North Carolina, Bohemia and elsewhere, alluvial diamonds were found in association with decomposed eruptive peridotite, which indicated serpentine as the original matrix of the diamond.[11]

On June 1, 1899, Reverend Professor Thomas George Bonney (1833–1923), a Vice President of the Royal Society and English geologist, followed in Lewis's footsteps with a paper about the parent rock of South African diamond.[12] By 1898, works at Kimberley had descended 450 metres, the last 35 metres of which was as hard as limestone. At that time, geologists shared more than one opinion about the origin of the diamonds in kimberlite. One thread in these varied opinions was that diamonds had crystallized in the matrix, perhaps by the action of molten lava on carbonate rock, or of steam in volcanic vents. The new suggestion put forward by Bonney was that a large crystalline mass had shattered in a deep explosion that had scattered grains of many minerals throughout the kimberlite. Bonney's samples were eclogite boulders recovered in 1897 at the Newlands Mine, some 60 kilometres northwest of Kimberley. When he examined their surfaces and slices of their interiors with a hand lens, he concluded that the diamondiferous rock was not the result of concretion of "blue ground." Bonney's important conclusion was that kimberlite was not the birthplace of the diamond.

> Thus, the diamond has been traced up to an igneous rock. The "blue ground" is not the birthplace of it or of the garnets, pyroxenes, olivine and other minerals ... which it incorporates. The diamond is a constituent of the eclogite. .... Its regular form suggests ... it was the first mineral to crystallise in the magma.

Bonney thought diamond was a derivative mineral that had precipitated out of molten magma.

⊙ CECIL RHODES: FOUNDER OF THE MODERN DIAMOND INDUSTRY

The commercial exploitation of African diamonds was swiftly taken under the control of capitalists rather than Fellows of the Geological Society. In 1871, Cecil John Rhodes (1853–1902) first saw the chaotic landscape of Kimberley's diamond fields when he was 18 years of age. He concluded that a new way of mining was urgently needed. Within two decades, he dominated the global trade and banked an immense fortune as a result of his outstanding political acumen and visionary dealmaking. One of nine sons of an Anglican vicar, and a sickly child, Rhodes' family dispatched him at the age of 17 to the much better climate of Natal for health reasons. In Natal, he worked on the cotton plantation that a brother had started, and when that venture failed, the two trekked 650 kilometres north to try their luck in the diamond fields. There Rhodes proved his entrepreneurial skills by selling ice to miners. With the profits from this activity, the future capitalist bought out individual diggers and recruited laborers to work his claims. And, like William Smith who had learned about the geology of coal from Somerset miners, Rhodes took advice from prospectors when forming his own view of the geological circumstances of Kimberley diamonds. He attended the sorting tables, carefully selecting and grading diamonds.

The open cast pits clustered atop a circular volcanic pipe about 500 metres wide. The friable yellow earth at the top of the pipe amounted to about 15 metres, below which the diamond-bearing rock became darker, finally taking on a slate blue or dark green color akin to serpentine. Solo prospectors were using their picks, buckets, ropes and pulleys to work four distinct diamondiferous pipes (Figure 14.1). Once all of the decomposed yellow ground on the surface had been shoveled into barrels and hauled aloft, the happy-go-lucky phase of diamond mining ended. Deeper mining was risky: hazardous collapses, fatal accidents and drownings in flooded workings ensued. The blue ground of unaltered rock was much harder to work, and the every-man-for-himself approach was hopelessly inadequate for recovering the ore with its concealed caches of diamonds. Throughout the 1870s, Rhodes aggressively snapped up claims as the exhausted pioneers pulled out. He invested capital in mining machinery and drainage. By 1874, he had solved the technical problems of pumping out the water flooding the deeper workings. He mobilized haulage machinery to hoist diamondiferous ore to the surface, where it dried out under the scorching sun. By 1880, he had

Mine de diamant à Kimberley

FIGURE 14.1 Postcard from the 1870s, illustrating the conditions at the Kimberley Mine around the period when Cecil Rhodes started to buy out the exhausted pioneers.
Source: Simon Mitton

consolidated his holdings to establish the De Beers Mining Company. Meantime, a rival entrepreneur, Barney Barnato (1851–1897), was also consolidating his holdings, such that by 1885 he had won control of the Kimberley Mine. Within a further three years, Rhodes, assisted by his business partner Alfred Beit (1853–1906), bought out Barnato and added Kimberley Mine to his portfolio. Cecil Rhodes now controlled nine-tenths of the world's trade in diamond gemstones.

On April 20, 1885, *The Times* of London carried a 4000-word dispatch from Kimberley that is remarkably vivid in its description of the operation of the mines, as well as the challenges facing mine-owners of thefts committed by their hired labor, "one of the great curses of the industry."

> Early in the digging the geologist stepped in to point out that these circular basins were evidently a species of volcanic crater, hollowed out in the surface by subterranean origin and filled up to the surface with a blue diamondiferous mud.

From a historical perspective, this extract is of interest because it informed a general readership in London that the Kimberley diamonds had been lifted from the deep by subterranean forces associated with volcanic activity. This is the earliest mention in the popular press that diamonds are xenoliths.[13]

⊙ KIMBERLITE AND THE FORMATION OF DIAMOND

Although Lewis coined the term "kimberlite" in 1887, the earliest mention in the literature of this new kind of rock goes back half a century to 1837. That's when Lardner Vanuxem (1792–1848) recorded the presence of igneous dykes of a kind that are kimberlitic at Ludlowville, near to Ithaca in New York State. Vanuxem had studied geology and mineralogy at the École des Mines, and two years after his graduation Vanuxem received an invitation from the University of South Carolina to fill their new chair of geology. He is credited with the introduction of stratigraphic paleontology to the USA.[14] The connection of diamond with kimberlite dates back almost 150 years. Curiously, though, our knowledge today of the circumstances of kimberlite eruptions remains quite modest. There are some five to six thousand known kimberlites, of which maybe a sixth are diamondiferous. Globally, there are about 50 commercial diamond mines. Kimberlites are hybrid rocks that cannot be described or classified by simple petrological models or by the presence or absence of particular minerals. They contain minerals, rock fragments and magmatic materials that formed in a variety of physical and chemical environments. Kimberlite is a highly alkaline ultramafic rock of igneous origin, usually found in carrot-shaped pipes (diatremes) 1–2 kilometres deep (Figure 14.2). The presence of diamond is indicative of its high-pressure origin, deeper than for any other rock. Eruptions of ultrabasic kimberlite have never been witnessed, so there is not really a consensus model of the eruptive mechanism. A publicly available database lists the locations of 4200+ kimberlites, with ages ranging from 45 Ma to 1.8 Ga.[15] In 2010, Trond Torsvik of the University of Oslo collaborated with colleagues at the University of the

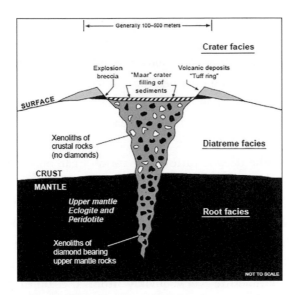

FIGURE 14.2 Schematic diagram of the anatomy of a kimberlite pipe present in an extinct volcano. (Note that the vertical scale is necessarily greatly reduced compared to the horizontal scale.) The explosive ascent of kimberlite magma follows a buildup of pressure from carbon dioxide and water. The kimberlite rises at speeds of hundreds of metres per second, entraining xenoliths including diamond. Rapid transfer of diamond from the lower mantle to the surface is essential because this carbon allotrope is unstable in the conditions of the upper mantle.

Source: Colorado Geological Survey, Golden, Colorado

Witwatersrand to examine this database and others for imprints of recent Earth history. Because kimberlites are derived from great depths, they provide invaluable information on the history and stability of the subcontinental mantle. By using plate reconstructions and deep tomography, they inferred that most kimberlite eruptions in the past 200 million years and possibly as far back as 540 million years are associated with two mantle plume generation zones at the core–mantle boundary. This would account for the concentration of kimberlites in continents with old cratons (ancient stable crystalline blocks forming the nuclear mass of a continent). Kimberlites are only known within continents, and most eruptions in the past 320 million years occurred in the part of the continent above a plume generation zone.[16]

Robert J. Stern (University of Texas at Dallas) and collaborators challenged Torsvik's interpretation. Using the same data, they developed a model in which an explosive kimberlite eruption could be triggered by the start of modern-style plate tectonics about a billion years ago (1 Ga).[17] They pointed out that subduction zones are our planet's largest recycling system. As oceanic lithosphere sinks in subduction zones, it can take down an enormous amount of H_2O and CO_2 to the deep mantle in the form of carbonate-rich and water-rich surface materials, plunging them to 670 kilometres or even deeper. It likely requires hundreds of millions of years for the water to be released from the descending slab and to rise to the base of the lithosphere where, under the right circumstances, it can set off an explosive kimberlite eruption. We know that diamonds have been transferred by kimberlite very quickly from the upper mantle (~150–240 kilometres) to the surface. If the transfer from the high-temperature, high-pressure regime of the mantle to the surface is protracted, unstable diamond is at risk of its carbon atoms being rearranged from a face-centered cubic crystal to the slippery hex-agonal layered structure of graphite. Ascending kimberlite shoots aloft at speeds in the tens of kilometres per hour, fast-tracking entrained diamond to the surface intact.

⊙ DATING DIAMONDS

Radiometric dating techniques are important for tracking the history of the mantle. Diamonds play an essential role in this because of the information locked in their inclusions. Decoding the messages avail-able within the inclusions has made a massive contribution to our understanding of the history and properties of the deep Earth. Gemologists reject diamonds with disfiguring inclusions because they cannot be transformed by the diamond cutters into attractive gems. These discards are eagerly sought by geologists because imperfections such as cracks, air pockets and foreign minerals such as garnet are birthmarks that tell us about the extreme and seemingly miraculous conditions in which the diamond crystallized in the mantle. Like a passport, the inclusions record the dates and times of the diamond's journey from the mantle to the surface.

Stephen H. Richardson (University of Cape Town) and Steven B. Shirey (Carnegie Institution of Science [CIS]) are among the pioneers who have drilled down to unearth the deep history of diamond formation. Almost four decades have passed since Richardson and three collaborators first published a landmark paper on the origin of diamonds in the remote past.[18] That study used samarium–neodymium (Sm–Nd) and rubidium–strontium (Rb–Sr) radiometric dating of garnets encapsulated in diamonds from relatively young (90 Ma) kimberlites from southern Africa. This technique is critically dependent on the sample being unaltered by fluids or weathering since its formation. A garnet locked in a diamond is therefore an ideal specimen. However, the garnets are very small, at ~100 microns, with a mass of ~10 micrograms. Richardson's team cracked numerous diamonds and, using a binocular microscope, they selected several hundred to supply nanograms of strontium and neodymium for analysis, from which they determined ages of 3.2–3.3 Ga: "Diamonds are forever!" Richardson acknowledged that "We are heavily indebted to De Beers Consolidated Mines for the provision of all durable raw materials."

Several years later, the Richardson team published the results of a more extensive examination of garnet inclusions from the Premier Mine, located 40 kilometres east of Pretoria.[19] This mine has been much in the news since 2010 because Petra Diamonds, the current owner–operator, has retrieved several large diamonds worth US$10–35 million each. Back in January 1905, the surface manager of the mine had famously picked up the 3106-carat Cullinan diamond, which remains the largest rough diamond ever found. Richardson and colleagues found that the samarium and neodymium signatures in the first generation of Premier diamonds are consistent with an Archean age (3.1–3.2 Ga). The second generation had an age of 1.93 Ga, while the third generation's age was 1.15 Ga. This research revealed changes in the composition and properties of the mantle across the three generations, thus yielding a further contribution to Earth history.

Together with colleagues at Carnegie and elsewhere, Shirey has developed a rhenium–osmium (Rh–Os) diamond dating procedure. The ^{187}Re isotope of rhenium decays to ^{187}Os very slowly indeed: the half-life is 41.6 Ga, which is good for measuring diamond ages up to 3.5 billion years. Shirey's methodology allowed researchers to date individual inclusions in a diamond, a marked advantage on the modeling approach that averaged the results from hundreds of diamonds. A remarkable conclusion from his research on deep diamonds sourced from southern Africa is that something extraordinary happened in Earth's history about three billion years ago: the onset of plate tectonics.[20] In 1968, Tuzo Wilson had identified a cyclical geodynamic process through which the global jigsaw puzzle of continental pieces dispersed and then reassembled on the Earth's surface, with ocean basins closing, continents colliding and mountain building. Shirey and Richardson found striking evidence of this process across a period of 3.5 billion years. By mining data in the literature on silicate and sulfide inclusions from five ancient cratons, they found that diamonds older than 3.2 Ga have inclusions with a peridotitic composition (ultramafic, olivine-rich), whereas inclusions younger than 3.2 Ga have an eclogitic composition (mafic, with green pyroxene and garnet). Such inclusions trapped inside natural diamonds are the most valuable *direct samples* we have of the deep mantle in the state that pertained billions of years ago. The suggested mechanism responsible for the change in the mineral inclusions of the diamonds from peridotitic to eclogitic was the geochemical evolution of our planet's interior following the onset of plate tectonics and subduction at 3 Ga, which marked the beginning of deep recycling of continental crust to the lower mantle.

⊙ ORIGIN OF THE LARGEST GEM DIAMONDS

Carbon in Earth occurs bound with oxygen as carbon dioxide or bound with hydrogen in methane and other organic molecules. In the first decade of the present century, strong evidence emerged that the formation of diamond requires the reduction of carbon in minerals to the bare elemental form. In the mantle, most of the diamond crystallizes from carbon-rich fluids (carbon–oxygen–hydrogen–sulfur), whereas in metamorphic rocks and meteorites, diamond is mainly the product of impacts.[21] However, extraordinarily beautiful diamond gems require extraordinary conditions of formation, and they are therefore of considerable scientific interest. The Cullinan diamond has long been regarded as having unusual characteristics of exceptional clarity and color, quite different from the diamonds recovered on an industrial scale in the most productive mines.

The centenary of the Cullinan find in 2005 seemed to herald a new era of diamond discovery, because by pure coincidence the Cullinan Mine has recently yielded the following gems: the 507-carat Cullinan Heritage (2009); a 25-carat blue diamond (2013); the 29-carat Blue Moon of Josephine and the 122-carat blue Cullinan Dream (2014); a 232-carat white diamond (2014); and 101-carat and 425-carat white diamonds (both recovered March 2019). Polishing these natural diamonds into fabulous gems is a lengthy process that produces a large number of offcuts to eliminate the disfiguring metallic and mineral inclusions. These offcuts with inclusions and impurities are much sought after for research on the origin of diamond.

Scientists classify diamonds according to the type and concentration of their chemical impurities. Type I diamonds have nitrogen as their principal impurity, typically at a concentration of 0.1 percent. About 98 percent of gem diamonds are colorless Type Ia, and in 0.1 percent of natural diamonds the nitrogen atoms are dispersed throughout the crystal in isolated clusters, giving the stone an intense yellow color: these rarities are classed as Type Ib. Type IIa diamonds have no nitrogen impurities and rarely show color: they are recognized by the different manner in which they absorb in the infrared range and transmit in the ultraviolet range, where they are very transparent down to wavelengths of 230 nm. Type IIb diamonds are rare and very valuable gems: they have the same very low level of nitrogen as Type IIa diamonds, but they contain significant boron impurities, which lends them a light blue or gray color.

The physical characteristics of the huge diamonds listed above posed some questions: How and where did they form? To solve these enigmas, Evan M. Smith of the Gemological Institute of America (GIA), which is appropriately located in New York's fabled Diamond District, teamed up with colleagues at the Department of Terrestrial Magnetism and the Geophysical Laboratory of the CIS, as well as at the University of Padova, Italy.[22,23] They followed up clues that massive gem diamonds are almost all Type IIa and devoid of nitrogen. The absence of nitrogen in Type IIa diamonds posed a major intellectual question. Plate tectonic activity provides a plausible mechanism for nitrogen to be recycled from the atmosphere into the mantle, where it then becomes a common ingredient in the fluids from which Type I diamonds form. With the resources of the diamond grading service at the GIA, Smith and colleagues sifted hundreds of thousands of polished gems and offcuts in search of a very specific subtype of diamond, to which they gave the inelegant acronym CLIPPIR, to signify "Cullinan-like, Large, Inclusion-Poor, Pure, Irregular and Resorbed." They sought diamonds in this category with tiny black metallic inclusions that were identified by their visual appearance. Just 53 of the selected diamonds had the specific inclusion assemblage that differentiated them from diamonds with flat black features typical of graphite. These diamonds had fractures surrounding many of the inclusions, a signature of the high stresses that build up as the diamonds are exhumed from the mantle in kimberlite eruptions.

The most common inclusion had a metallic composition, which at the temperature and pressure of diamond formation would have been an iron–nickel–carbon–sulfur mixture. The level of detail regarding the conditions in the mantle that was achieved by this research is fascinating: "These inclusions are accompanied by a thin fluid layer of methane and hydrogen trapped at the interface between the solid inclusion and surrounding diamond."[22] Detailed analysis revealed that the minerals that crystallized in the small droplets of molten metallic liquid were trapped in the diamond as it ascended to the surface. The researchers concluded that the best and biggest

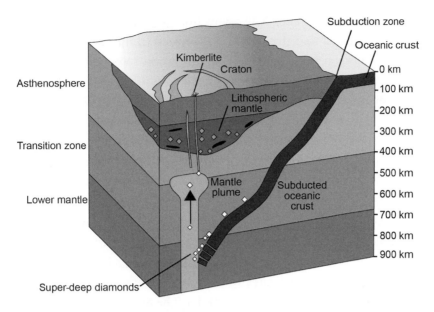

FIGURE 14.3 Rare super-deep diamonds are formed from subducted oceanic crust at depths of 400–700 kilometres. They have a much deeper origin than that of most gem diamonds, which originated in the lower part of continental tectonic plates at 150–200 kilometres. Highly prized blue diamonds owe their color to the element boron, which was conveyed to the deep mantle by subduction.

Source: Penny Wieser

diamonds formed in metallic liquid at extreme depths (~410–660 kilometres) in the convecting mantle (Figure 14.3). This is a much deeper origin than that of most gem diamonds, which originated in the lower part of continental tectonic plates at 150–200 kilometres. A key component of the research is that the mantle below 250 ± 30 kilometres can precipitate iron–nickel metal to the point of saturation, and then the dissolution of carbon and sulfur into this metal lowers its temperature below the liquidus, so it becomes fluid. For diamond to crystallize, there must be sufficient carbon to reach supersaturation.

Seven of the CLIPPIR diamonds were also suitable for investigation of their carbon isotopic composition at CIS. These seven had deviations from the composition of the standard that indicated a deficit in light carbon ($\delta^{13}C$) varying from –26.9 to –3.8 per mil. This result suggested that the carbon that formed CLIPPIR diamonds was recycled carbon from the crust! Confirmation soon came when the Smith collaboration performed a similar investigation of 46 Type IIb blue diamonds with inclusions; these constitute ≤0.02 percent of mined diamonds.[24] As mentioned above, their blue color arises from the presence of boron at a level of 0.01–10 parts per million. Boron is a lithophile: it is abundant in the crust, but with a low concentration in the mantle. So, the formation conditions for blue diamonds had long been deeply puzzling. The 46 boron-bearing diamonds were found to carry previously unrecognized mineral assemblages formed under high pressure in the lower mantle. In their 2018 paper in *Nature*, the authors suggested that boron dissolved in seawater is transported in hydrous magnesium silicates to the lower mantle, where it is liberated in a hydrous diamond-forming fluid. The resulting blue diamonds are eventually lifted to the surface by mantle upwelling followed by kimberlite volcanism, as is the case for Type IIa diamonds. The research on super-deep diamonds conducted by the GIA and CIS in 2016–2018 not only solved the enigma of the source of the world's biggest diamonds, but also ushered in a new frontier for deep carbon science: the laboratory investigation of the extreme physics and chemistry of the deep mantle by examining high-pressure mineral inclusions inside diamonds that were scooped up in kimberlite and thence propelled explosively to the surface.

REFERENCES

1. Shirey, S. B. and Shigley, J. E. Recent advances in understanding the geology of diamonds. *Gems and Gemology* **49**, 188–222 (2013).
2. Godard, G. and Chabou, M. C. First African diamonds discovered in Algeria by the ancient Arabo-Berbers: history and insight into the source rocks. *Comptes Rendus Geoscience* **346**, 179–189 (2014).
3. Streeter, E. W. *The Great Diamonds of the World: Their History and Romance* (G. Bell & Sons, 1882).

4. Walker, E. A. *South Africa, Rhodesia and the High Commission Territories*. **8** (Cambridge University Press, 1963).

5. *Treaty of Lahore* (1849).

6. Broun-Ramsay, J. Personal letter to Queen Victoria (1874).

7. Dunn, E. J. On the mode of occurrence of diamonds in South Africa. *Quarterly Journal of the Geological Society* **30**, 54–60 (1874).

8. Story-Maskelyne, N. and Flight, N. On the character of the diamantiferous rock of South Africa. *Quarterly Journal of the Geological Society* **30**, 406–416 (1874).

9. Beasley, A. W. *Dunn, Edward John (1844–1937)*. Australian Dictionary of Biography (Melbourne University Press, 1981).

10. Lewis, H. C. On a diamantiferous peridotite and the genesis of the diamond. *Geological Magazine* **4**, 22–24 (1887).

11. Lewis, H. C. The matrix of the diamond. *Geological Magazine* **5**, 129–131 (1888).

12. Bonney, T. G. The parent-rock of the diamond in South Africa. *Proceedings of the Royal Society of London* **65**, 223–236 (1900).

13. Diamonds wholesale. *The Times*, **4** (1885).

14. Johnson, M. E. *Vanuxem, Lardner (1792–1848)*. American National Biography (Oxford University Press, 1999).

15. Faure, S. World kimberlites CONSOREM database (version 3). In *Consortium de Recherche en Exploration Minérale CONSOREM* (Université du Québec à Montréal, 2010).

16. Torsvik, T. H., Burke, K., Steinberger, B., Webb, S. J. and Ashwal, L. D. Diamonds sampled by plumes from the core–mantle boundary. *Nature* **466**, 352–355 (2010).

17. Stern, R. J., Leybourne, M. I. and Tsujimori, T. Kimberlites and the start of plate tectonics. *Geology* **44**, 799–802 (2016).

18. Richardson, S. H., Gurney, J. J., Erlank, A. J. and Harris, J. Origin of diamonds in old enriched mantle. *Nature* **310**, 198–202 (1984).

19. Richardson, S. H., Harris, J. W. and Gurney, J. J. Three generations of diamonds from the old continental mantle. *Nature* **366**, 256–258 (1993).

20. Shirey, S. B. and Richardson, S. H. Start of the Wilson cycle at 3 Ga shown by diamonds from subcontinental mantle. *Science* **333**, 434–436 (2011).

21. Chen, M., Shu, J., Xie, X., Tan, D. and Mao, H. K. Natural diamond formation by self-redox of ferromagnesian carbonate. *Proceedings of the National Academy of Sciences* **115**, 2676–2680 (2018).

22. Smith, E. M. *et al.* Large gem diamonds from metallic liquid in Earth's deep mantle. *Science* **354**, 1403–1405 (2016).

23. Smith, E. M., Shirey, S. B. and Wang, W. The very deep origin of the world's biggest diamonds. *Gems and Gemology* **53** (2017).

24. Smith, E. M. *et al.* Blue boron-bearing diamonds from Earth's lower mantle. *Nature* **560**, 84–87 (2018).

15 Deep Life

⊙ SIGNS OF DEEP LIFE

This chapter on deep carbon subsurface life opens at the 2018 Fall Meeting of the American Geophysical Union (AGU) in Washington, DC, where Deep Carbon Observatory (DCO) scientists showcased stupendous discoveries about deep life.[1] Earth's most pristine ecosystem, the deep biosphere, is home to members of all three domains of life: Archaea, Bacteria and Eukarya.[2,3] Archaea and Bacteria are microbes, and the Eukarya include fungi, algae, unicellar organisms with organelles, as well as plants and animals. Unicellular organisms exist everywhere on Earth's surface, from the thermophiles in the hot springs of Yellowstone National Park to the microbes living in your refrigerator or below the ice sheets of Siberia and Antarctica. The huge surprise that captivated the public following the press releases at AGU was the immense mass of carbon directly associated with subsurface bacterial life. Researchers estimated that this reservoir holds 15–23 billion tonnes of organic deep carbon.

The history of research on deep life starts in the recent past, but we have been actively engaging with microbes and yeasts for thousands of years to produce bread, beer, wine and cheese. Scientific pursuit of deep subsurface life began in 1790, thanks to the Prussian naturalist and polymath Alexander von Humboldt (Figure 5.6). During 1791–1792, young Alexander was enrolled at the Freiberg Academy of Mines, where he followed an intensive curriculum of geology and engineering. On his field trips to the pits of Saxony, he became fascinated by the mosses, fungi and algae that flourished underground in the dank darkness. On graduation, he was appointed to a Prussian government position as an inspector of mines, which opened up frequent opportunities to study the subterranean cryptogams in the gold mines of the Fichtel Mountains of Bavaria. Humboldt's position today as a late eighteenth-century pioneer of deep life rests on this extensive research, published in 1793.[4]

In the mid-nineteenth century, the French chemist Pierre Jacques Jean Antoine Béchamp (1816–1908) made the surprising discovery that rock itself can harbor life.[5] No geologist or biologist had ever suggested that solid rock could be an abode of life. Béchamp was born in the ancient rural hamlet of Bassing in Lorraine. During his teenage years, a kindly uncle who worked as an assistant in the French diplomatic service raised him in Romania. Following his uncle's passing in 1834, he trained in pharmacy at Strasbourg, where he earned two doctorates, and succeeded Louis Pasteur (1822–1895) as Professor of Chemistry in 1854. Béchamp enters our story through a key paper published in 1868. By then, of course, the microscopical works of Robert Hooke and Antonie van Leeuwenhoek (1632–1723) were well known, the latter having shown that microbes (anaerobes) can live without oxygen. Béchamp published a startling finding: the crushed chalk he used to maintain neutrality in acid fermentations in milk contained not only tiny fossils of crustaceans, but also very small organisms, smaller than yeasts, which were still alive despite their extreme geological age.[6] They grew rapidly in the laboratory, nourished by a variety of organic substances.[7] He put it thus:

> We have evidence that microzyma [enzymes] of chalk are not
> specific ferments [yeasts]; in general there is none; what is there are
> organisms that provoke or transform the nature of the food they are
> provided with.

This was the first suggestion that organisms can exist inside rock and survive on *chemical* energy.

Tullis Onstott, a geomicrobiologist at Princeton University, has expressed a modern take on the thriving endolithic life lurking in rock:

We do not normally think of rock as harboring life ... I would wager
that the last thought in your mind as you gaze on Michalangelo's
David ... in Florence is the fact that within microscopic pores
buried inches beneath the surface of smooth Carrara marble are
living bacteria. These bacteria may have been trapped inside the
marble for a million years and are slowly reproducing, releasing
carbon dioxide ...[8]

It was only in the 1920s that a suspicion of the presence of
microbes in the deep subsurface – hundreds of metres below ground –
took hold. In Illinois, methane had been occasionally observed in
water wells, where it posed an explosion hazard. The director of the
state water survey attributed the source of the dissolved gas to the
organic decay of vegetable matter buried in glacial drift, deposited
over a time span of 600,000 years.[9,10] Also in Illinois, Edson
Sunderland Bastin (1878–1953), professor of economic geology at the
University of Chicago, reported that sulfate-reducing bacteria were
present "in abundance with some of the waters associated with oil in
productive fields." The oil-producing horizons from which the
samples were obtained were at depths of 100–400 metres. Bastin
remarked that some geologists regarded the oil-field waters "as
ancient sea waters buried within the rocks." He also speculated that
it might never be possible to tell if the bacteria in sediments had been
laid down 300 million years ago or were introduced much later by
seepage of ground waters to the oil-bearing horizons below.[11] In 1930,
Bastin's lab succeeded in growing bacteria from oil-well water drawn
at depths of 275–600 metres. This signaled that a subsurface biosphere
might be present, but at that time there was little enthusiasm for deep
microbial life, given the difficulty of being certain that samples from
the oily sediments had not been contaminated.

Serious interest in the science of deep microbial life was revived
in the 1980s, when the US Department of Energy (DOE) became
concerned with identifying safe storage facilities for nuclear waste.
Vast quantities of radioactive trash were being simply dumped at

industrial facilities and were contaminating groundwater. DOE scientists were urgently searching for underground repositories that could safely house high-level radioactive waste for thousands of years. Frank J. Wobber, a scientist and manager at the DOE, approached these environmental issues by recognizing that the fundamental science about deep aquifers was nonexistent. He supposed that if microbes were present deep below the surface, they might either benignly degrade organic pollutants or dangerously disrupt the chambers being used for the long-term storage of radioactive waste. Either way, fundamental research on deep microbial life had become a priority, so in 1985, Wobber initiated the DOE Subsurface Science Program. He brought together geologists, biologists and chemists in the search for deep-dwelling life. The Program hit the ground in 1987, when the DOE arranged for the drilling of several boreholes in South Carolina, near to its huge Savannah River Plant, a nuclear processing facility. An interdisciplinary team of researchers from more than 20 universities and laboratories worked together on this project, led by Brent F. Russell. They examined the abundance and diversity of the microbes in the sediments, carrying out applied research directed at the biological restoration of deep subsurface environments contaminated with nuclear waste.

Their emphasis was on securing pristine samples from the Savannah River deep aquifers. Chemical and physical tracers were used in the drilling fluids to check for microbial contamination.[12] These samples were shipped immediately to research laboratories across North America specifically for investigation of their diverse communities of microbes. By 1989, Wobber, Tommy Joe Phelps (Oak Ridge National Laboratory) and others had documented the discovery of diverse communities of bacteria and archaea at a depth of 500 metres.[13] Those accustomed to dismissing such claims finally became persuaded when it was found that the subsurface microbes could not survive in the presence of oxygen. Furthermore, they were found to have unusual genetic sequences. The Savannah River project marked a turning point for the acceptance of a deep biosphere that had been isolated from the atmosphere for at least 100 million years. Microbial populations in the deep sediments were more abundant and more active than had been expected, and an extensive number of bacteria new to science were isolated.[14]

The discoveries of deep life made by investigators of groundwater contamination now set off a global revolution in geomicrobiology. To circumvent the prohibitive expense of drilling to the deep biosphere, biologists and geologists teamed up with the companies drilling exploratory wells. That's what DOE scientists did to study rocks from the Taylorsville Basin in eastern Virginia. They broke the then-current record for well-documented subsurface life in 1993, when they collected a new species of bacteria from 2.7 kilometres below the surface. These were thermophiles – heat-loving bacteria – that earned the name *Bacillus infernus* for their hellish habitat. By measuring the size of pores in the host rock, the researchers established that the indigenous bacteria had been trapped in the rock for millions of years, where they were surviving on organic compounds dissolved in groundwater. In simple terms, these microbes were just "hanging out" without dividing (apparently), and had been doing so for 80–160 million years.[15]

⊙ DRILLING DOWN IN SWEDEN

The astrophysicist and space scientist Thomas Gold (1920–2004) was a singular scientist who delighted in crossing any field where he felt that the cult of "what is generally accepted" had led to other options being ignored. In a long and varied scientific life, he frequently stepped across boundaries, venturing into astronomy and cosmology, biophysics, aerospace engineering, planetary science and geophysics. Gold puzzled over philosophical questions on the origins of life in the universe and its possible habitats. He (and his colleague Fred Hoyle) eventually concluded that what geologists had been teaching about the *biological* origin of petroleum should be challenged. They advanced anthropic arguments, such as the following: "[G]iven that hydrocarbons exist in meteorites, and on other planets, and in interstellar space, it seems very strange to argue that other planets and interstellar space have hydrocarbons of nonbiological origin, whereas Earth has obtained them from a different source ... namely biology." By following this logic, Gold concluded that most of Earth's hydrocarbons must be *primordial* in origin. Gold's abiotic hypothesis was that since carbon, carbon dioxide, methane and carbonates exist in the mantle, the hydrocarbons found in petroleum could have been generated from these raw materials at high temperatures and pressures. The resulting fluids, being less dense than rock, would percolate upward over geological time through cracks and fissures. And since the geologists were never going to waste their time on a hypothesis that they deemed absurd, Gold set about finding evidence of such in his typically independent manner.[16]

He launched a highly effective campaign to drill for deep hydrocarbons in a location that had never been probed by the oil industry: Sweden, a land of largely bare granite with scarcely any sedimentary rock. Central Sweden hosts a very large meteorite impact crater 45 kilometres in diameter: the Siljan Ring. This was of prime interest to Gold because the impact would have smashed through the crust, thereby generating pathways to a great depth and filling much of the crater with porous crushed rock. In the 1980s, the Swedish State Power Board (Vattenfall) desperately needed to find new sources of energy because Swedish voters had declared in a 1980 referendum that the country's 12 nuclear power plants must be shut down by 2010. After initial hesitation, Vattenfall reacted favorably to Gold's hypothesis on deep hydrocarbons, and a geochemical soil survey was undertaken. This confirmed that petroleum indicators were as high as in any petroleum-bearing area in the USA. Five exploratory holes were then drilled to a depth of about 500 metres. This was deemed a success in terms of demonstrating the possible presence of hydrogen and methane.

Despite the incredulity of exploration geoscientists, the Vattenfall energy company, together with corporate and private investors, raised US$40 million to fund unconventional drilling through granite to a depth of 7.2 kilometres, where the seismic profiling had indicated a porous zone. Two boreholes were drilled, one to 6.7 kilometres and a second to 6.5 kilometres – it was impossible to go deeper. Expert geochemists checked the meager samples of hydrocarbons recovered and came to the conclusion that diesel oil and additives in the drilling fluids had been altered by the frictional heating of the drill bit. There were no natural hydrocarbon reservoirs suitable for commercial operations. Gold, of course, disagreed, and he continued to believe in deep microbial life, even without having found compelling evidence at the Siljan Ring. The expense of drilling through the crust for samples and the time required suggested that a different approach was needed in the search for subterranean life.

⊙ SUBTERRANEAN LIFE IN DEEP MINES

For more than two decades, Tullis Onstott's research has been driven by arduous fieldwork on endolithic deep life that has taken him to the deepest gold mines of South Africa and North America. After completing his doctorate in geology at Princeton in 1980, he spent three years as a postdoctoral fellow at the University of Toronto working on argon isotope geochronology. Back at Princeton, as an early-career scientist, he applied $^{40}Ar/^{39}Ar$ geochronology and spectroscopy in a wide range of investigations: fieldwork in South Africa on the gold fields and kimberlites, dating of diamonds and studies in paleomagnetism. In 1992, he attended a meeting at the DOE headquarters in Maryland, where Terry Phelps was speaking on anaerobic thermophiles. At the end of the meeting, Frank Wobber stressed the importance of developing dating methods for subsurface microbial communities: Were they Triassic or much more recent? *Jurassic Park* flashed through Onstott's mind: Was it possible that ancient living organisms were still viable deep beneath Earth's surface?[8] Deep microbial life intrigued him. His new agenda became a "search for deep and potentially very ancient microbial systems supported by radiation and to search for the origins of life itself."[8]

Building on his experience of the South African mines, Onstott secured an invitation to search for microbial life at Mponeng in the Western Deep Levels southwest of Johannesburg in 1996. It is currently the world's deepest mine. The miners guided him to its deepest tunnels, where he grabbed a large chunk of radioactive black rock (carbon leader ore) and sealed it in a steel canister filled with argon gas. He checked in his radioactive rock as baggage on his flight home. The only sign of life discovered was a species of bacterium that had previously been identified from a hot spring in Mexico. The surprise was that this new species could reduce (transfer electrons to) iron and uranium, with a consequential release of energy. In 1998, Onstott and his team visited the Dreifontein Mine, where they collected fracture water samples, which are much easier to handle than rock. These samples of ancient seawater had bacterial populations similar to those that were then being found at hotspots and vents in the deep seafloor (described below in the "Deep Marine Life at Hydrothermal Vents" section). By the turn of the century, it was abundantly clear that deep rocks in South Africa's gold mines had flourishing communities of archaea and bacteria wherever there was a supply of water.

To explore the deep subsurface directly, Onstott put together a multidisciplinary team that would carry out fieldwork in the world's deepest accessible excavations. One recruit was the early-career microbiologist Ken Takai, a postdoctoral visitor from Japan. He set to work on extracting DNA from water drawn from boreholes and fissures. Sequence and phylogenetic analyses were then used to provide data that would characterize the different communities and their environments. Novel archaeal sequences revealed by ribosomal DNA (rDNA) analyses suggested that the gold mines and their underlying strata were home to distinctive microbial communities. An intriguing find was that rDNA clones from fissure water in the Kloof Mine "may have been derived from a *Pyrococcus* species living in aerobic saline groundwater at a depth of 5 to 6 km." This was the first suggestion that hyperthermophiles (already identified at *seafloor* vents) might exist in the deep *terrestrial* subsurface.[17]

Onstott's project attracted Barbara Sherwood Lollar (University of Toronto), one of Canada's most renowned and highly honored geoscientists. Her application of isotopic geochemistry to subsurface water and light hydrocarbons had already been transformative in the fields of environmental science and subsurface life. In 1993, Sherwood Lollar and her colleagues had investigated methane discharges from fractures and hard-rock boreholes in mines located across the Canadian Shield. They noticed that the $\delta^{13}C$ and $\delta^{2}H$ signatures were incompatible with methane derived from organic matter, but had some similarities to those from the Murchison meteorite. This indicated that rock–water interactions were the source of the hydrocarbon gases.[18] Confirmation of this hypothesis came in 2002 from the demonstration that hydrocarbon gases from the Kidd Creek (Ontario) mine had $\delta^{2}H$ values consistent with hydrogen derived from radiolysis, serpentinization and polymerization reactions.[19] In 2006, Sherwood Lollar, Onstott and their colleagues demonstrated the role of such hydrogen-rich fluids in sustaining microbial communities at >2.8 kilometres below the surface at Mponeng.[20] Over a period of 54 days, they sampled fracture water four times to determine whether contamination had occurred. DNA microarray analysis revealed a microbial biome dominated by a single new species of sulfate reducer belonging to the phylum Firmicutes. Isotopic analysis of the fracture water indicated a minimum age of ~3–4 Ma. In a press release accompanying the publication of these results, Onstott enthused:

> These bacteria have been cut off from the surface of the Earth for many millions of years but have survived in conditions most organisms would consider as inhospitable to life. Could these bacterial communities sustain themselves no matter what happened on the surface? If so, it raises the possibility that organisms could survive even on planets whose surfaces have long since become lifeless.[21]

⊙ DEEP MARINE SEDIMENT COMMUNITIES

One year then passed before microbiologists Todd O. Stevens and James P. McKinley of the Pacific Northwest National Laboratory located bacteria living off bare rock 1.5 kilometres beneath the Columbia River Plateau, a large igneous province formed 10–15 million years ago. This was a stunning first: an active anaerobic microbial ecosystem in deep basalt aquifers deriving its energy from geochemically produced hydrogen in the interstices of basalt flows 3–5 kilometres thick.[22] Conversely, other researchers soon challenged the suggestion that the microbial population was supported by hydrogen: they argued that the hydrogen detected in basalt microcosms must have been produced by water–rock reactions involving ferrous silicate minerals.[23] Stevens and McKinley issued a rebuttal in 2000 when they defended their hypothesis that the deep microbial ecosystems were supported by the abiotic production of hydrogen.[24] Then the pendulum swung back the other way when a population analysis of two deep communities located within the Columbia River Basalt Group found that fewer than 3 percent of its microbes were methanogens.[25]

The ongoing debate took an interesting turn when astrobiologists considered the possibility of deep subsurface life on Mars. Arguing that a subsurface habitat on Mars was unlikely to contain organic matter in sufficient quantities to support bacterial communities, they suggested instead that the interaction of hydrothermal fluids with igneous rocks could release hydrogen as an energy source. Researchers with the United States Geological Survey (USGS) followed up this idea with fieldwork in Idaho looking for microbial ecosystems that lacked organic matter but did have a source of geologically produced hydrogen. At Lidy Hot Springs, Idaho, 200 metres below the land surface, Francis Chapelle and colleagues discovered "a unique microbial community in which hydrogen-consuming, methane-producing archaea far outnumber the bacteria." The geochemical character of the community indicated that geothermal hydrogen, not organic matter, was the source of energy, all of which was consistent with the microbial habitats that might exist in the subsurface of Mars.[26]

⊙ MARINE BACTERIA AND BIOFILMS

In two seminal papers from the 1930s, one by Arthur T. Henrici (1889–1943), professor of bacteriology at the University of Minnesota,[27] and a second by Claude ZoBell (1904–1989) drew attention to communities of microbes growing as films on submerged surfaces.[28] Henrici and ZoBell were working on the ecology of marine bacteria in an era when most of their peers studied specimens cultivated within the laboratory rather than observing the bacteria in their marine habitats. Henrici urged "a movement away from the time-honored techniques of bacteriology toward a more rational study of the microbe in its natural environment." He described the "new branch of science, ecology" as "essentially a revolt from the ... methods of the laboratory," and he called for "a return to the aims of an old-fashioned field naturalist." Henrici had become curious about the vigorous growth of algae in his aquarium. His curiosity intensified when he found that microscope slides immersed in the aquarium for a week acquired a filmy deposit of bacteria, adhering so firmly that it could not be dislodged by washing under a tap. Henrici repeated the experiment with water drawn from several sources: other aquaria, the university's lily pond and a large lake: "In every case the results were the same." His conclusion was that "it is quite evident that for the most part water bacteria are not free-floating organisms but grow attached upon submerged surfaces."

ZoBell was in charge of microbiology at Scripps, where he sampled bacteria found in bottom deposits at depths of up to 1300 metres. His analysis of marine sediments showed "the existence of a large and varied microflora," but attempts to reveal the presence of bacteria in water at depths of greater than 500 metres had failed.[29] He and Esther Allen (a visiting doctoral student from Scripps) were also working on the fouling of submerged structures. It had been known since antiquity that barnacles on ship hulls are a big drag, and reports commissioned by the US Navy in 1923 had suggested that the formation of a slime film preceded the unwelcome arrival of barnacles. In February and March of 1933, ZoBell and Allen ran a series of quantitative trials to study the organisms attached to submerged glass slides. Their most noteworthy result was that 2-square-inch slides submerged for 48 hours acquired bacteria by the million, whereas the seawater had only hundreds of bacteria per cubic centimetre. They concluded that bacteria are the primary film-formers, and that bacterial film provides a surface that promotes the attachment of macroscopic organisms. In the 1930s, this was "pioneering science" because it set bacteriology on new pathways in natural science and marine science that were distinctive from those of health science and medical science in which most bacteriologists had worked hitherto. By the end of the twentieth century, biologists understood what biofilm communities implied for microbial ecosystems. It was widely accepted that "bacteria cells have the ability to aggregate into particular three-dimensional assemblages, differentiate and hence divide labor within these assemblages, and then disperse as part of their life cycle."[30]

⊙ DREDGING AND DRILLING THE DEEP

Early in 1950, ZoBell and Richard Y. Morita (1949–2005) participated in a joint Scripps and US Navy mid-Pacific expedition examining the seafloor between San Diego and the Marshall Islands. This was the first Scripps deep sea expedition with the research vessel *Horizon*, and its primary purpose was to acquire bottom sediment from the deep seafloor: they obtained 75 cores of up to 8 metres in length.[31] In red clay and planktonic ooze, ZoBell and his doctoral student Morita found bacterial populations ranging from hundreds to thousands per gram of pelagic sediment. The abundance of viable bacteria decreased with core depth. Living bacteria found at a depth of 8 metres were in a state of "slow and continuous activity" in sediments more than one million years old.[32]

The pair next embarked the corvette *Galathea*, an unarmed Royal Danish Navy research vessel. ZoBell led the microbiology

FIGURE 15.1 Marine microbiologist Claude ZoBell preparing a water sampling bottle in 1952 on board *Galathea*.
Source: Scripps Institution of Oceanography, University of California, San Diego

investigations of this expedition: sediment samples were hauled from depths of 7000–10,180 metres, the latter then being a new record for sampling deep life (Figure 15.1). More than 20,000 people lined the quay to welcome *Galathea* back on June 29, 1952: throughout this voyage, accompanying journalists and film crews had run a superb outreach mission. ZoBell spent a couple of years investigating the cultural conditions for successful incubation in the laboratory: pressures of 70 MPa (700 atm) or more were required for any signs of life from the "barophilic" bacteria.[33] Despite ZoBell's pioneering research on the deep biosphere, which had found microbes wherever sediments were dredged up, agnostics remained suspicious of contamination.

⊙ DEEP MARINE LIFE AT HYDROTHERMAL VENTS

In 1964, the US Navy commissioned *Alvin*, a manned deep-ocean research submersible the size of a small truck, which was seconded to the National Oceanic and Atmospheric Administration (NOAA) and operated by the Woods Hole Oceanographic Institute (WHOI). It remains in operation, and the science payback from its thousands of research dives has been absolutely immense (Figure 15.2). In the 1970s, a highly productive theme of *Alvin* expeditions was the

FIGURE 15.2 The US Navy Deep Submersible Vessel *Alvin* entering service at Woods Hole, Massachusetts, on June 5, 1964, prior to being placed in service.
Source: National Archives, Washington, DC

investigation of thermal hotspots on the seafloor associated with mid-ocean ridge spreading centers. The science driver behind this research was the tectonic model of seafloor spreading. Researchers anticipated that hydrothermal systems should offer plenty of opportunities for rapid chemical reactions and new science. In the east Pacific, there is a triple junction of three plates: the Pacific, Cocos and Nazca plates to the west of the coastlines of Ecuador and Colombia.[34] Early in 1972, John Sclater and Kim Klitford made a detailed survey of the heat flow anomalies at this spreading center, concluding that the variability of the heating and cooling of near-bottom seawater was caused by hydro-thermal circulation through small fissures in recently intruded bas-altic magma.[35] They speculated that "the observed temperature anomalies are more probably caused by thermal plumes arising from hydrothermal vents."

The next part of the story on the search for hydrothermal vents is drawn Robert Ballard's compelling account published in *Oceanus*.[36] By early 1977, oceanographers were fully prepared for a major geophysical research effort in the east Pacific. On February 8, 1977, the Woods Hole research vessel RV *Knorr* nosed out of the Panama Canal en route for the deep Galápagos Rift valley, a yawning gap of an oceanic spreading ridge between the Cocos and Nazca plates. There were geologists, geochemists and geophysicists aboard, but no one had anticipated that there might be any role for a marine biologist. On February 15, the deep-towed camera sled *Angus* was deployed portside to image the deep seafloor. In a 12-hour shift, *Angus* would glide 4.5 metres above the seafloor, its camera rig and powerful strobe lights imaging the seafloor every 10 seconds. Every photograph had a precise time stamp in its lower left corner. Crucially, *Angus* also had the ability to detect changes in temperature as small as 0.005°C. Suddenly, at about midnight, the temperature probe sensed a spike in the water temperature that had lasted three minutes. This anomaly was not quite a "wow!" moment, but it was a thrilling wake-up call for the oceanographers. The next day, the research team scrutinized the photographs frame by frame. The lava landscape imaged just seconds before the temperature anomaly had a fresh-looking appearance. But in the 13 frames corresponding to the spike in temperature, the lava was hidden by hundreds of large white clam and mussel shells, a startling accumulation of deep marine life not glimpsed before.

An epic dive in *Alvin* ensued at sunrise on February 17, 1977, and the team undertook two dozen dives in the following six weeks. This series enabled direct visual observations of the area of the temperature anomaly, the taking of physical measurements to quantify the fluxes and thermodynamics of the rock–water interactions occurring in the hydrothermal system and the taking of samples of fluids and deposits on the seafloor. Ballard participated in this dive series, on which he commented later:

> But when they reached their target coordinates, *Alvin* and its three-man crew entered another world. Coming out of small cracks cutting across the lava terrain was warm, shimmering water that

quickly turned a cloudy blue as manganese and other chemicals in solution began to precipitate out of the warm water and were deposited on the lava surface, where they formed a brown stain. But even more interesting was the presence of a dense biological community living in and around the active vents. The animals were large, particularly the white clams (up to 30 centimeters or 12 inches). This basis of life was only 50 meters across and totally different from that of the surrounding area. What were the organisms eating? They were living on solid rock in total darkness.

Alvin dipped water samples (50–100 litres) from four hot vents. The laboratory analysis of these pointed the way to answering the question: How can these deep-sea creatures flourish in total darkness with no obvious signs of nutrition? When the first sample was opened, "[T]he smell of rotten eggs filled the lab." Ugh! The water was saturated with hydrogen sulfide. Portholes were unlatched to clear the foul-smelling gas. John Edmond (1943–2001), one of the geochemists, later said that halfway into the cruise they began "to realize regular seawater was mixing with something – it was a unique solution, never seen before." The hydrogen sulfide seemed to be the key to understanding deep life sustained by the fluids streaming out of the hydrothermal vents. Strong interactions were occurring between the water and basalt at temperatures of at least 300°C. Jack Corliss and 10 coauthors of these interdisciplinary studies concluded that "these fragile communities provide a unique opportunity for a wide range of zoological, bacteriological, ecological, and biochemical studies."[37] Needless to say, the biologists were bursting with enthusiasm to investigate these strange oases of extraordinary life.

David Perlman (1918–2020), a renowned science journalist at the *San Francisco Chronical*, was a guest on the expedition, with his Olivetti portable typewriter at the ready. He telecopied up-to-the-minute reports to the paper, which heralded Perlman's biggest scoop under the banner headline "Astounding Undersea Discoveries":

They have pinpointed geysers of hot water venting from fissures in fresh lava and sending warm plumes of brine shimmering upward into the near-freezing lower levels of the sea ... They have found rich clusters of living organisms, basking in the warmth of the geysers ... They have discovered fresh lava that was poured out onto the sea bottom in ropes and wrinkles, sheet-like pavements and bulbous pillows – squeezed or erupted from the hot, semi-molten material of the deep earth's interior mantle beneath the crust. When these findings are all analyzed in detail they are bound to revolutionize many theories about the deep ocean floor.[38]

Pursuing that "analysis in detail" consumed US$200,000 for the design and fitting out of many upgrades to the sampling equipment on *Alvin*. The big scientific conundrum was the food source of the "rich clusters of living organisms." Holger W. Jannasch (1927–1998) of WHOI conducted experiments in which he filled syringes containing radioactive $^{14}CO_2$ with vent fluids and incubated them for two days. This biochemical investigation established that vent bacteria absorb hydrogen sulfide and oxidize it to sulfur. Very unexpectedly, microbial sulfur oxidation was found to be a key deep marine process leading to dense animal communities. A vital aspect of the discovery was the close interdisciplinary collaboration between biologists, chemists and geophysicists.

The ecosystem at the deep ocean hot springs is absolutely extraordinary, energized by the heat of Earth, with bacteria thriving in a superhot environment saturated with toxic chemicals and heavy metals that would be lethal for most life on the continents. No wonder that marine biologist John Frederick Grassle (1939–2018) and the WHOI chief scientist on the 1979 diving expedition later wrote, "Nothing could diminish the excitement of seeing the animals for the first time."[39] In his autobiography, Jannasch recalled:

We were struck by the thought, and its fundamental implications, that here solar energy, which is so prevalent in running life on our planet, appears to be largely replaced by terrestrial energy, liberated by the oxidation of reduced inorganic chemicals, with chemolithoautotrophic bacteria taking over the role of green plants. This was a powerful new concept and, in my mind, one of the major biological discoveries of the twentieth century. Communicating it to the geochemists was a delightful learning experience.[40]

For a dive on April 21, 1979, scientific observers Bill Normark (1943–2008) of the USGS and Thierry Juteau, a French volcanologist, were piloted by Dudley Foster, who followed a trail of white clam shells on the seafloor, hoping it was a pointer to a hydrothermal vent. Suddenly, an incredible structure about 2 metres high sprang into view, spilling out a black cloud of fluid. When *Alvin* bumped into this chimney, it disintegrated, revealing a wide vent billowing with blackness. As Foster manipulated the robotic arm to shove a temperature probe into the heart of darkness, the reading rocketed to the upper limit of the thermometer: 32.7°C. He assumed that the thermometer was acting up, but when an engineer later checked what was wrong with the temperature probe, he was shocked to see that its plastic tip had melted. The melting point of that type of plastic is 180°C! The viewports on *Alvin* were made of the same plastic: in the near-freezing (~2°C) depths, they had ventured to within a couple of metres of searing hot brine at 350°C. Hydrothermal fluids react with crustal rocks by dissolving metals, which precipitate as sulfides as the superheated water hits very cold, oxygen-rich seawater. The astonishing discoveries of 1977 and 1979 precipitated the final acceptance by Congress to fund crewed submersibles as an essential tool for worldwide oceanographic exploration by the USA.[41]

By the end of the century, more than 200 vent fields had been logged in the ocean basins; their associated metal deposits and their biota of clams, tube worms and shrimp supported by microbial activity were well documented. These high-temperature systems were found along the very narrow axis (~5 kilometres) of mid-ocean ridges where 85 percent of the world's output of magma is localized. At the same time, research papers on life in extreme environments increasingly referred to organisms found in extreme physical conditions. Astrobiologists passionately lauded the extremophiles when speaking of habitats for life elsewhere in the solar system, and biologists interested in the origin of life on Earth regarded the hydrothermal vents and hot springs as "doorways into early Earth."[42,43] More recently, Onstott has commented that when "the first communities of organisms were discovered at black-smoker vents at the bottom of the Pacific Ocean ... no one had imagined that such complex organisms [tube worms] and microbial communities could survive on chemical gradients in the utter darkness of the seafloor."[8]

Another great leap forward for oceanographic science occurred on December 4, 2000, when a cruise by a new US Navy research vessel, RV *Atlantis*, to the Mid-Atlantic Ridge serendipitously made a spectacular find at the majestic Atlantis Massif soaring 4250 metres from the seafloor, the same height as Mount Rainier. The Massif is composed of ancient dense green rock (peridotite) rather than young black basalt. On board RV *Atlantis*, chief scientist Donna K. Blackmann (Scripps) and her colleagues Deborah S. Kelley (University of Washington) and Jeffrey A. Karson (Duke University) were noting the detailed structure, composition and dynamic history of the Atlantis Massif in a bid to understand how this giant block of peridotite and gabbro had found its way to the surface. When they scanned the cliffs of the Massif with a high-resolution camera system, amazing white structures loomed into view, seemingly made of coral (Figure 15.3). It soon became apparent to petrologist Gretchen Früh-Green (ETH Zürich) and Kelley that they had discovered a fantastic cluster of chimneys towering over hydrothermal vents.[44] In the busy schedule for *Alvin*, only one dive was possible for getting samples. When Deborah Kelley looked carefully at the collection baskets, she spied tiny organisms wrenched from the fragile chimneys.

A return expedition in April–May 2003 consisted of 19 dives of *Alvin* and a further 17 surveys by a robotic underwater vehicle. Many questions needed answers because the Lost City complex (so named by its discoverers on *Atlantis*) with its towering carbonate monoliths

FIGURE 15.3 Lost City Hydrothermal Field. A 1.5-metre-wide flange, or
ledge, on the side of a chimney is topped with dendritic carbonate growths
that form when mineral-rich vent fluids seep through the flange and come
into contact with the cold seawater.
Source: University of Washington; Woods Hole Oceanographic Institution

was unlike any hydrothermal system seen before. The active geochemistry at the Lost City Hydrothermal Field (LCHF) was entirely
new, being based on serpentinization reactions between seawater and
mantle peridotite. Exothermic reactions were driving fluids rich in
hydrogen and methane out of the chimney vents, which were then
precipitating carbonates on mixing with seawater. This was providing
the abundant thermophilic and lithophilic organisms with plenty of
nutrients.[45] When Früh-Green, Kelley and their coinvestigators
looked at the mineral forms in the chimneys and used strontium,
carbon and oxygen isotope data and radiocarbon ages to construct a
history of the LCHF and the Atlantis Massif, they concluded that
hydrothermal activity at the LCHF driven by serpentinization had
extended over at least 30,000 years. But that was just a lower age
limit: "[T]he amount of heat produced by serpentinization has the
potential ... to drive Lost City-type systems for hundreds of thousands if not millions of years." Whereas hydrothermal vents
centered on an ocean ridge become inactive once the spreading
seafloor has moved away from the magmatic heat source, the
longevity at the LCHF could be far greater because the heat source –
serpentinization – could in theory endure for millions of years,

while continuously pumping out methane for the sustenance of microbial vent communities.[46] Given the range and complexity of environments that host peridotite, the research impact of the findings at the LCHF was vast: the links between serpentinization, the formation of carbonates and the microbial activity now had enormous implications for the origin and sustainability of deep life on Earth. Wherever peridotites and other ultramafic rocks were wetted with briny fluids, hydrothermal systems could spring to life! Could deep life exist anywhere in the ocean basins, not just on the oceanic ridges? Was this how life arose on Earth?

⊙ THOMAS GOLD PROPOSES THE DEEP,
      HOT BIOSPHERE

The discovery of abundant life at hydrothermal vents established that life could exist independently of photosynthesis by using chemical energy from the reaction of seawater with rock. In 1992, in a paper titled "The deep, hot biosphere," Thomas Gold posed the following question:

> How widespread is life based on internal sources of the Earth? Are the ocean vents the sole representatives of [deep life], or do they merely represent the examples that were discovered first? ... We may well expect that other locations that are harder to investigate would have escaped detection so far.[47]

He noted that many research groups were describing the discovery of thermophilic bacteria in deep locations where they had not been anticipated, and that suggested to him a simple principle: "that microbial life exists in all locations where microbes can survive." His definition of survival invoked the availability of chemical energy and an ambient temperature high enough to activate the reactions but less than ~100°C. Gold was fully aware that in our solar system bodies, only Earth has favorable conditions for *surface* life. Elsewhere, astrobiologists were thinking of *subsurface* life being the norm. On the emergence and evolution of life, Gold speculated:

> The surface life on the Earth, based on photosynthesis ... may be just one strange branch of life, an adaptation specific to a planet that happened to have ... favorable circumstances on its surface as would occur only very rarely [in the universe].

It was then but a small step for him to add that deep, chemically supplied life may be very common in the universe. In his posthumous memoir (2012), he writes that his premise then was:

> Within the Earth's crust, there exists a second biosphere, composed of very primitive heat-loving bacteria and containing perhaps more living matter than is present on the Earth's entire surface.

Gold's independent thinking on a deep biosphere provoked the skeptics and naysayers, but at the same time it sparked keen interest among geomicrobiologists, with the extent and mass of the under-world soon becoming big issues. In a nutshell: How much is down there? The discovery of deep marine life described above was certainly unexpectedly diverse, but it was not deep *subsurface* marine life. So far, all attempts to detect deep life in the LCHF have been unsuccessful, possibly due to the extremely high pH (>12), which may be too hostile for subsurface life to tolerate.

⊙ DRILLING TO THE DEEP BIOSPHERE

Back in the late 1980s, microbiologists first discovered living micro-organisms in sediment cores obtained through the Ocean Drilling Program, and in the following decade, observations were made at drill sites in the Atlantic and Pacific oceans, as well as the Mediterranean Sea. In 1994, Ronald John Parkes (University of Bristol) and eight colleagues retrieved marine sediments on Ocean Drilling Program cruises to five Pacific Ocean sites. In the deep biosphere, they discovered viable populations of microbes buried under more than 500 metres of sediment that appeared to be living on ancient organic matter.[48] By combining data on cell numbers from these sites, they found a systematic 1000-fold decrease from the sediment surface

(~10^9 cells cm^{-3}) to several hundred metres below the seafloor (~10^6 cells cm^{-3}), where the sediment had been deposited many million years ago.[49] Intriguingly, the researchers' finding that the populations did not decrease with increasing depth encouraged the intriguing thought that it is likely that bacterial populations are present at much greater depths. By extrapolation, they suggested that "global sedimentary organic carbon, remarkably is 10% of the living carbon in the surface biosphere." The authors felt that "the deep biosphere responds to ... geochemical fluxes ... probably more significant globally than biological interactions at hydrothermal vents and in oil reservoirs."

By 1998, sufficient data on the subsurface cell densities of archaea and bacteria had accumulated to enable three microbiologists at the University of Georgia to publish the first global inventory of the total amount of cellular carbon on Earth; William B. Whitman was the lead author.[50] They reached the astonishing conclusion that the subsurface biomass in the deep seabed more than equaled the biomass in the deep terrestrial subsurface. The deep biosphere appeared to account for one-third of the biomass of the entire planet! However, they conceded that their assessments had been poorly constrained by sparse data, cautioning that "future studies could revise this view and the subsurface biomass may be overestimated." That seemed to be the case in 2000 when John Parkes *et al.* opted for a marine subsurface biosphere at one-tenth of the world's biomass. Despite the reduction, this was still a staggering amount. By this stage of research, the ubiquity of deep life in sub-seafloor sediments had been established beyond reasonable doubt, but little was known of its diversity, the composition of its communities and its variation from one environment to another. It was time to drill deeper to get answers.

In January 2002, a two-month expedition of the research drilling vessel *JOIDES Resolution* (ODP 201) took a shipboard scientific party, led by marine oceanographer Steven D'Hondt of the University of Rhode Island, to seven sites in the equatorial Pacific Ocean and continental margin of Peru. ODP 201 was the first ever drilling expedition to focus on microbiology. The water depth varied from a shallow continental Peru shelf to a vertiginous plunge of 5300 metres in the Peru Trench. Sediments up to 35 million years in age were sampled. Bacteria were successfully cultured and isolated from multiple depths at every site, indicating that living bacteria were present throughout the entire range of sub-seafloor depths, with the deepest samples being from 420 metres below the seafloor. Most of the microorganisms have no cultured or known relatives in the surface world, and at the time of their discovery they were only characterized by the genetic code of their DNA.[51] Their metabolic activities unexpectedly exhibited great diversity. Concentration profiles of dissolved chemicals indicated that biologically mediated reactions were consuming and releasing metabolites deep in the sediment column at all sites. There, biogeochemical evidence of activity included organic carbon oxidation, methanogenesis, sulfate reduction and manganese reduction. At open-ocean sites, the data clearly indicated that microbes in the sediment were not being fueled by thermogenic CH_4 from deep within Earth. Buried organic matter from the photosynthetic world appeared to be sufficiently abundant to fuel organic methanogenesis.

The findings that ODP 201 made in 2002 had a major impact on our understanding of microbial activity below the seafloor, and it provided a clear agenda for future microbial investigation: reach the greatest depth that sub-seafloor organisms attain; refine knowledge of their effects on global biogeochemical cycles; and determine the total genetic diversity, rate of population turnover and community structures.[52] Fifteen years later, a citation analysis and retrospective review of Gold's deep, hot biosphere proposal concluded that it had

... inspired a generation of researchers in the field of geobiology to dive deeper into the possibilities of subsurface life, spawning hundreds of relevant publications. Importantly, considerable funding from the [National Science Foundation] for drilling expeditions [since 2006], in concert with the Integrated Ocean

Drilling Program efforts, have led to tremendous new insights in marine subsurface settings.[53]

For Expedition 329 of the Integrated Ocean Drilling Program (IODP 329) to the South Pacific Gyre (SPG), RV *Roger Revelle* departed Samoa on December 17, 2006, with a group of scientists drawn from several institutions with very different areas of expertise. This cruise explored sediments that have accumulated extraordinarily slowly over tens of millions of years (0.1–1 metre per million years). The central SPG is Earth's largest oceanic desert, and it contains the clearest seawater in the world. SPG sediments are further from continents and productive ocean zones than any other site on Earth. Incidentally, the area of the SPG is more than twice the area of North America. The purpose of IODP 329 was to research the nature of deep life locked in the most oxidized and food-limited sediments of Earth's oceans. At every depth in cored sediments, the microbiologists measured mean cell abundances that were three to four orders of magnitude lower than at the same depth in all previously explored sub-seafloor communities. Despite the extremely low rate of organic burial, the SPG biomass has subsisted in the sediment for at least 70 million years. For these communities, the generation of H_2 by the radiolysis of water caused by radiation from radioactive decay is a significant electron-donor source.[54]

The striking difference of several orders of magnitude between cell counts derived from probes of the ocean margins and east equatorial Pacific Ocean on the one hand and of oceanic gyres on the other hand prompted an assessment in 2012 of the variation of microbial abundance throughout the world and its effects on estimates of the sub-seafloor sedimentary biomass and the total biomass.[55] Jens Kallmeyer and four colleagues compiled their cell counts from the two oceanic gyres with previously published counts from the ocean margins. Not surprisingly, they found a strong correlation between the organic burial rate and sub-seafloor sedimentary cell abundance: a classic example of supply-and-demand equilibrium economics. Also, it is no surprise, perhaps, that with a database far more geographically diverse than those of previous studies because it included the very sparse counts of the gyres, their estimate of the sub-seafloor sediment biosphere was less than one-tenth (92 percent lower) the estimate of Whitman *et al.* in their 1998 paper.

⊙ EPILOGUE

This history of the pioneering research on deep life is no more than a brief summary of half a century of extraordinary progress on questions about the limits of life on Earth and the nature of deep life. For me, as an astrophysicist by training, the highlight of my intellectual journey has been the extraordinary revolution in microbiological science in the last half-century. For much of my academic career, I have been writing about remarkable developments in astronomy and cosmology: the discovery of the high-energy universe, its dark energy and its dark matter – an intellectual exercise in "looking up and looking out." But in the last four years, I have been "looking down and looking deep" at Earth's interior. I have spent my entire academic career at the University of Cambridge, so I was already familiar with the history of the great discoveries by its astronomers and physicists, as well as its geologists and geophysicists. In 2002, I cofounded the interdisciplinary *International Journal of Astrobiology* as a forum for practitioners in cosmic prebiotic chemistry, the search for Earth-like exoplanets, Mars as an abode of life and extremophile biology. From my present-state examination of the research on deep life, I have reached a philosophical conclusion that there is a great deal still to be discovered about the microbial environment of Earth's interior that is highly relevant to the search for life elsewhere. When it became clear that today there is no surface life on Mars, the nature of the deep life on Earth began to feature prominently in astrobiological fora. Therefore, it seems entirely plausible to me that, right beneath our feet, deep carbon science could yet uncover important clues on the origin of life in the universe.

REFERENCES

1. Magnabosco, C. *et al.* The biomass and diversity of the continental subsurface. *Nature Geoscience* **11**, 707–717 (2018).

2. Woese, C. R. and Fox, G. E. Phylogenetic structure of the prokaryotic domain: the primary kingdoms. *Proceedings of the National Academy of Sciences* **74**, 5088–5090 (1977).

3. Woese, C. R., Kandler, O. and Wheelis, M. L. Towards a natural system of organisms: proposal for the domains Archaea, Bacteria, and Eucarya. *Proceedings of the National Academy of Sciences* **87**, 4576–4579 (1990).

4. von Humboldt, A. *Florae Fribergensis specimen plantas cryptogramicus prae-sertim subterraneas exhibens* (Berolini, H.A. Rottmann, 1793).

5. Béchamp, A. Lettre de M. A. Béchamp à M. Dumas. *Annales de chimie et de physique* **13**, 103–111 (1868).

6. Béchamp, A. Du rôle de la craie dans les fermentations butyrique et lactique, et des organismes actuellement vivants qu'elle contient. *Comptes Rendus* **63**, 451–455 (1866).

7. Manchester, K. L. Antoine Béchamp: père de la biologie. Oui ou non. *Endeavour* **25**, 68–73 (2001).

8. Onstott, T. C. *Deep Life: The Hunt for the Hidden Biology of Earth, Mars, and Beyond.* **1** (Princeton University Press, 2017).

9. Buswell, A. M. and Larson, T. E. Methane in ground waters. *Journal (American Water Works Association)* **29**, 1978–1982 (1937).

10. Wickham, J. T. *Glacial Geology of North-Central and Western Champaign County, Illinois* (llinois Institute of Natural Resources, 1979).

11. Bastin, E. S. The presence of sulphate reducing bacteria in oil field waters. *Science* **63**, 21–24 (1926).

12. Russell, B. F., Phelps, T. J., Griffin, W. T. and Sargent, K. A. Procedures for sampling deep subsurface microbial communities in unconsolidated sediments. *Groundwater Monitoring & Remediation* **12**, 96–104 (1992).

13. Ghiorse, W. and Wobber, F. J. Special issue on deep subsurface microbiology – introduction. *Geomicrobiology Journal* **7**, 1–2 (1989).

14. Sargent, K. A. and Fliermans, C. B. Geology and hydrology of the deep subsurface microbiology sampling sites at the Savannah River Plant, South Carolina. *Geomicrobiology Journal* **7**, 3–13 (1989).

15. Monastersky, R. Deep dwellers: microbes thrive far below ground. *Science News* **151**, 192–193 (1997).

16. Gold, T. *Taking the Back Off the Watch: A Personal Memoir.* **381** (Springer Science & Business Media, 2012).

17. Takai, K. E. N., Moser, D. P., DeFlaun, M., Onstott, T. C. and Fredrickson, J. K. Archaeal diversity in waters from deep South African gold mines. *Applied Environmental Microbiology* **67**, 5750–5760 (2001).

18. Lollar, B. S. *et al.* Abiogenic methanogenesis in crystalline rocks. *Geochimica et Cosmochimica Acta* **57**, 5087–5097 (1993).

19. Lollar, B. S., Westgate, T. D., Ward, J. A., Slater, G. F. and Lacrampe-Couloume, G. Abiogenic formation of alkanes in the Earth's crust as a minor source for global hydrocarbon reservoirs. *Nature* **416**, 522–524 (2002).

20. Lin, L. H. *et al.* Long-term sustainability of a high-energy, low-diversity crustal biome. *Science* **314**, 479–482 (2006).

21. These bacteria use radiated water as food. *Indiana University Press Release* (2006).

22. Stevens, T. O. and McKinley, J. P. Lithoautotrophic microbial ecosystems in deep basalt aquifers. *Science* **270**, 450–455 (1995).

23. Anderson, R. T., Chapelle, F. H. and Lovley, D. R. Evidence against hydrogen-based microbial ecosystems in basalt aquifers. *Science* **281**, 976–977 (1998).

24. Stevens, T. O. and McKinley, J. P. Abiotic controls on H_2 production from basalt–water reactions and implications for aquifer biogeochemistry. *Environmental Science & Technology* **34**, 826–831 (2000).

25. Fry, N. K., Fredrickson, J. K., Fishbain, S., Wagner, M. and Stahl, D. A. Population structure of microbial communities associated with two deep, anaerobic, alkaline aquifers. *Applied Environmental Microbiology* **63**, 1498–1504 (1997).

26. Chapelle, F. H. *et al.* A hydrogen-based subsurface microbial community dominated by methanogens. *Nature* **415**, 312–315 (2002).

27. Henrici, A. T. A direct microscopic technique. *Journal of Bacteriology* **25**, 277–287 (1933).

28. ZoBell, C. E. and Allen, E. C. The significance of marine bacteria in the fouling of submerged surface. *Journal of Bacteriology* **29**, 239–251 (1935).

29. *Report of the Scripps Institution of Oceanography* (Scripps Institution, 1934).

30. Davey, M. E. and O'Toole, G. A. Microbial biofilms: from ecology to molecular genetics. *Microbiology and Molecular Biology Reviews* **6**, 847–867 (2000).

31. Revelle, R. *Director's Report on the Mid-Pacific Expedition of the University of California and the U. S. Navy Electronics Laboratory. UC San Diego Library, Special Collections. Scripps Institution of Oceanography Letters, Clippings and Ships' Logs* (University of California San Diego, 1950).

32. Morita, R. Y. and ZoBell, C. E. Occurrence of bacteria in pelagic sediments collected during the mid-Pacific expedition. *Deep Sea Research* **3**, 66–73 (1953).

33. ZoBell, C. E. and Morita, R. Y. Barophilic bacteria in some deep sea sediments. *Journal of Bacteriology* **73**, 563–568 (1957).

34. McKenzie, D. P. and Morgan, W. J. Evolution of triple junctions. *Nature* **224**, 125–133 (1969).

35. Sclater, J. G. and Klitgord, K. D. A detailed heat flow, topographic, and magnetic survey across the Galápagos Spreading Centre. *Journal of Geophysical Research* **78**, 6951–6975 (1973).

36. Ballard, R. D. Notes on a major oceanographic find (marine animals near hotwater vents at ocean bottom). *Oceanus* **20**, 35–44 (1977).

37. Corliss, J. B. *et al.* Submarine thermal springs on the Galápagos Rift. *Science* **203**, 1073–1083 (1979).

38. Perlman, D. Astounding undersea discoveries. *San Francisco Chronicle* (1977).

39. Grassle, J. F. Biologists' first look at vent communities – Galápagos Rift. *Oceanus* **41**, 1–5 (1979).

40. Jannasch, H. W. Small is powerful: recollections of a microbiologist and oceanographer. *Annual Review of Microbiology* **51**, 1–45 (1997).

41. Ballard, R. D. The history of Woods Hole's Deep Submergence Program. In *50 Years of Ocean Discovery* (National Academy Press, 2000), pp. 67–84.

42. Rothschild, L. J. and Mancinelli, R. L. Life in extreme environments. *Nature* **409**, 1092–1101 (2001).

43. Beal, H. Life in Extremely Hot Environments. *Microbial Life Educational Resources*. Available at: https://serc.carleton.edu/microbelife/extreme/extremeheat/index.html (accessed June 24, 2019).

44. Basgall, M. *Descent to the Mid-Atlantic Ridge: Expedition Journals* (Regents of the University of California, 2000).

45. Kelley, D. S. *et al.* A serpentinite-hosted ecosystem: the Lost City Hydrothermal Field. *Science* **307**, 1428–1434 (2005).

46. Früh-Green, G. L. *et al.* 30,000 years of hydrothermal activity at the Lost City vent field. *Science* **301**, 495–498 (2003).

47. Gold, T. The deep, hot biosphere. *Proceedings of the National Academy of Sciences* **89**, 6045–6049 (1992).

48. Parkes, R. J. *et al.* Deep bacterial biosphere in Pacific Ocean sediments. *Nature* **371**, 410–413 (1994).

49. Jørgensen, B. B. Shrinking majority of the deep biosphere. *Proceedings of the National Academy of Sciences* **109**, 15976–15977 (2012).

50. Whitman, W. B., Coleman, D. C. and Wiebe, W. J. Prokaryotes: the unseen majority. *Proceedings of the National Academy of Sciences* **95**, 6578–6583 (1998).

51. Jørgensen, B. B. and D'Hondt, S. A starving majority deep beneath the seafloor. *Science* **314**, 932–934 (2006).

52. D'Hondt, S. *et al.* Distributions of microbial activities in deep subseafloor sediments. *Science* **306**, 2216–2221 (2004).

53. Colman, D. R., Proudel, S., Stamps, B. W., Boyd, E. S. and Spear, J. R. The deep, hot biosphere: twenty-five years of retrospection. *Proceedings of the National Academy of Sciences* **114**, 6895–6903 (2017).

54. D'Hondt, S. *et al.* Subseafloor sedimentary life in the South Pacific Gyre. *Proceedings of the National Academy of Sciences* **106**, 11651–11656 (2009).

55. Kallmeyer, J., Pockalny, R., Adhikari, R. R., Smith, D. C. and D'Hondt, S. Global distribution of microbial abundance and biomass in seafloor sediment. *Proceedings of the National Academy of Sciences* **109**, 16213–16216 (2012).

Glossary

Compiled by Fiona E. Iddon

abyssal plain. Deep ocean floor lying between the base of the continental rise and the mid-ocean ridge.

achondrite. Class of meteorite that does not contain chondrules. The parent bodies of achondrites have undergone differentiation.

alpha particle. Nucleus consisting of two protons and two neutrons bound together.

aragonite. Carbonate mineral, one of the three most common naturally occurring crystal forms of calcium carbonate.

asthenosphere. The denser, semifluid layer of hot plastic rock below the lithospheric mantle.

biofilm. A thin layer of mucilage adhering to a solid surface, containing the community of bacteria and other microorganisms that generated it.

black hole. An object that has collapsed to such a small volume that its immense gravitational field does not allow light to escape.

black smokers. Chimneys formed from iron sulfide deposits at hydrothermal vents.

carbonaceous chondrite. Class of meteorite with a primitive composition representative of the early solar system. Their chemical makeup includes organic compounds.

carbonation. Reactions of carbon dioxide to produce carbonates, bicarbonates and carbonic acid.

carbonic acid. Chemical compound with the formula H_2CO_3.

carboxylation. Chemical reaction in which carboxylic acid is produced by the addition of carbon dioxide to a substance.

chemosynthesis. Synthesis of organic compounds by living organisms using energy derived from reactions involving inorganic chemicals.

chondrite. Class of meteorite that contains chondrules – rounded grains of igneous origin – as a significant component of their composition.

continental drift. The theory that Earth's continents move relative to one another.

core. The innermost portion of a planet. Earth has a solid inner core and a liquid outer core, both composed primarily of dense, hot iron–nickel alloy.

crust. The thin, solid outer layer of a planet.

cryptogam. Plant that reproduces using spores.

Curie point . The temperature above which certain materials lose their permanent magnetic properties, to be replaced by induced magnetism.

diatreme . Volcanic pipe formed by a gaseous explosion, often when rising magma interacts with groundwater.

dwarf planet . Planetary-mass object that orbits a star but has not yet cleared the surrounding space of other material in its orbit.

echo sounding. Sonar hydrographic sensing technique used to determine the distance between the water surface and underwater topography by transmitting sound pulses.

eclogite . A high-temperature, high-pressure metamorphic rock of igneous origin, common in upper mantle regions.

electron microprobe. An instrument in which a sample is bombarded with an electron beam, causing the emission of X-rays at wavelengths characteristic of the elements being analyzed.

electron microscope. An instrument that uses a beam of accelerated electrons as the source of illumination.

Euler pole. Term in spherical geometry. The uniform motion of a rigid body on the surface of a sphere can be described as a rotation about a fixed axis of rotation or Euler pole.

extremophiles. Organisms that thrive in physically or geochemically extreme conditions.

fluorescence. The emission of light by a substance that has absorbed light or other electromagnetic radiation.

fracture zones. Major discontinuities disrupting the linearity of spreading mid-ocean ridge segments, they form depressions separating two regions of plates that slip in opposite directions.

fumarole. Opening in Earth's surface that emits steam and volcanic gases.

geomagnetic reversal. Change in a planet's magnetic field such that the positions of magnetic north and magnetic south are interchanged.

geosyncline. Late nineteenth-century concept: geosynclines were defined as giant downward folds in Earth's crust that served as a means to explain orogens.

giant molecular cloud. Dense interstellar cloud of gas and dust. Composed primarily of molecular hydrogen, they are important sites of star formation.

gneiss. High-grade regional metamorphic rock characterized by its foliations of alternating light and dark bands.

graben. Valley with a distinct escarpment on each side caused by the downward displacement of a block crust.

great circle. Circle on the surface of a sphere that lies in a plane passing through the sphere's center. It represents the shortest distance between any two points on a sphere.

half-life. The time required during radioactive decay for the amount of the parent element to reduce its initial value by half.

hydrogenation. Chemical reaction in which pairs of hydrogen atoms are added to generally unsaturated compounds.

hydrothermal vent. A fissure in Earth's surface providing an outlet for geothermally heated water.

isotope. Atoms of the same element that differ in the number of neutrons they contain.

kaolinite. A layered silicate clay mineral that forms as a result of chemical weathering of feldspar and other aluminum silicate minerals.

kimberlite. A highly alkaline ultramafic rock of igneous origin, usually found in diatremes, sometimes containing diamonds.

lias. Nineteenth-century English term for hard limestone and also the time span in which it formed.

lithophile. Microorganisms that live within the pore interstices of sedimentary and even fractured igneous rocks.

lithosphere. The rigid outermost layer of a terrestrial planet that is defined by its rigid mechanical properties.

magnetic variation. The angle on the horizontal plane between magnetic north and true north.

magnetite. An iron oxide mineral.

mantle. A layer within a terrestrial planet and some other rocky planetary bodies. For a mantle to form, the body must have undergone the process of planetary differentiation.

mass spectroscopy. Experimental technique used to identify the chemical isotopes in a substance by first bombarding it with electrons to create ions, which are then sorted according to the mass-to-charge ratio of the ions.

matrix. The finer-grained mass of material in which larger grains, crystals or clasts are embedded.

mid-oceanic ridge. A chain of underwater mountains marking a boundary between tectonic plates that are moving apart, with new oceanic crust being created here via decompression mantle melting.

nepheline syenite. A completely crystalline igneous intrusive rock consisting largely of nepheline, a silica-undersaturated aluminosilicate (feldspathoid) and alkali feldspar.

neptunism. Late eighteenth-century theory that rocks formed from crystallization of minerals in Earth's primeval oceans.

neutron star. Type of star with a mass up to 1.4 times that of the Sun; about 10 kilometres in diameter and made almost entirely of neutrons.

oceanic core complex. An uplifted dome of lower-crustal and upper-mantle rocks, the roof is stripped off by extreme tectonic extension.

oceanic trench. Topographic depressions in the seafloor that mark the position where a subducting slab begins to descend beneath another tectonic plate.

oolite. Form of limestone made up of ooids – spheroidal, coated sedimentary grains usually composed of calcium carbonate.

P wave. A compressional seismic body wave that shakes the ground back and forth parallel to the direction of wave travel.

paleomagnetism. Branch of geophysics concerned with the magnetism in rocks that was induced by Earth's magnetic field.

paleopole. Earth's magnetic pole as it was situated at a time in the distant past.

peridotite. Ultramafic igneous rock consisting primarily of olivine and pyroxene.

petrology. Branch of geology that involves the study of rocks and the conditions under which they form.

phosphorescence. The emission of radiation in a similar manner to fluorescence, but on longer timescales, even after excitation ceases.

photon. An individual quantum of light.

planetesimal. Small bodies of accreted gas and dust in orbit around a star, from which protoplanets and eventually planets are believed to have formed.

plate tectonics. The theory that explains the large-scale motion and interaction of rigid lithospheric plates.

pleochroic haloes. Spherical patterns of visible discoloration within some minerals caused by radiation damage.

polar wandering. The movement of Earth's poles relative to the continents throughout geological time.

porphyritic. An igneous rock texture in which larger distinct crystals are contained in a finer groundmass.

protoplanet. Large bodies of matter in orbit around a star that have undergone internal melting to produce a differentiated interior.

protoplanetary disk. The rotating disk of dust and gas surrounding a young star from which orbiting celestial bodies may eventually form.

radioactive decay scheme. The graphical representation of all of the transitions occurring during radioactive decay and their relationships.

Raman spectroscopy. Analytical technique based on inelastic scattering of monochromatic light. The frequency of photons in monochromatic light changes upon interaction with a sample.

red giant star. Luminous giant star of low or intermediate mass in the late phase of stellar evolution.

refractive index. The dimensionless number that describes how fast light propagates through a material.

S wave. The shear seismic body wave that shakes the ground back and forth perpendicular to the direction of wave travel.

seismic reflection/refraction. Techniques that measure the travel time of seismic energy from surficial shots through the subsurface to arrays of sensors. In the subsurface, seismic energy is refracted and/or reflected at interfaces between materials with different seismic velocities.

seismic time travel curve. A graph of the elapsed time for seismic waves to travel from the epicenter of an earthquake to seismic stations as a function of distance.

serpentinization. The low-grade hydration and metamorphic transformation of ultramafic rock.

siderophile. An "iron-loving" element that has little affinity for oxygen and sulfur.

stellar wind. The continuous flow of plasma ejected from the upper atmosphere of a star.

stereographic projection. A mapping function that projects a sphere onto a plane; the projection is defined on the entire sphere, except at the projection point.

stratigraphy. Branch of geology concerned with the order and relative position of rock layers.

strike and dip. The orientation or attitude of a planar geological feature, such as a rock bed or fault. The strike line represents the intersection of that feature with a horizontal plane, while the dip gives the steepest angle and direction of descent.

subduction. The movement of a tectonic plate beneath another plate.

supernova. A star that explodes at the end of its lifetime, either totally disrupting or leaving a condensed object, either a black hole or neutron star.

thermophile. An organism that can survive at high temperatures that are detrimental to most life on Earth.

thrust faulting. Also known as reverse dip slip faulting, it results from compressional forces caused by shortening of Earth's crust.

transform fault. A plate boundary where the motion is predominantly horizontal.

transmutable. The ability to change from one form, nature, substance or state to another.

trias. Nineteenth-century German term used for a series of terrestrial mud and sandstones. Also, the time span in which they formed.

triple-alpha process. The set of nuclear fusion reactions that transform three helium-4 nuclei into one carbon nucleus.

tuff. Light, porous rock formed by consolidation of volcanic ash.

unconformity. A hiatus in the sedimentary geological record, where the older rocks are generally exposed to erosion before deposition of the younger rock layer.

uniformitarianism. The theory that changes in Earth's crust over time are the result of continuous and uniform processes.

volatile elements. Chemical elements and compounds that have a low boiling point.

volcanic arc. Chain of volcanoes formed above a subducting plate.

xenolith. An area within an igneous rock that is not derived from the original magma but has been derived from elsewhere.

X-ray diffraction. A technique used for determining the atomic and molecular structure of a crystal.

Biographical Notes

Compiled by Iddon Fiona E.

Agricola, Georgius (Georg Bauer, Georg Pawer) (1494–1555)

German physician, mineralogist and metallurgist, known as the father of mineralogy. He published the first illustrated textbook on ore mineralogy.

Arrhenius, Svante (1859–1927)

Swedish physicist and chemist, Arrhenius was the first scientist to model the effect of atmospheric carbon dioxide on global warming.

Baade, Walter (1893–1960)

German–American astronomer who calibrated the cosmic distance scale and estimated the age of the universe as 4 billion years.

Bauer, Louis Agricola (1865–1932)

American geophysicist and astronomer, Bauer was the first director of the Department of Terrestrial Magnetism of the Carnegie Institution. He mapped the global magnetic field.

Béchamp, Pierre Jacques Jean Antoine (1816–1908)

French chemist, physician and pharmacist who discovered that organisms can exist in rock and thrive within them by using chemical energy.

Becquerel, Antoine Henri (1852–1908)

French physicist who made the serendipitous discovery of spontaneous radioactivity.

Berner, Robert Arbuckle (1935–2015)

American geologist noted for his contributions to modeling of the geochemical carbon cycle and its effect on atmospheric carbon dioxide over the past 100 million years.

Berzelius, Jons Jacob (1779–1848)

First chemist to make a distinction between organic and inorganic compounds. His study of the composition of the Alais meteorite is notable for the identification of water.

342   BIOGRAPHICAL NOTES

Biot, Jean-Baptiste (1774–1862)

French physicist, astronomer and mathematician, Biot is known for studying the
meteorites that fell on L'Aigle in 1803. The eyewitness reports he gathered
showed beyond doubt that the fallen stones were of extraterrestrial origin.

Blackett, Patrick (1887–1974)

British experimental physicist whose work on the physics of magnetism led him to
build a sensitive magnetometer that provided evidence of continental drift.

Bode, Johann (1747–1826)

German astronomer who reformulated and popularized the Titus–Bode law, which
attempted to explain the distances of planets from the Sun as a numerological
sequence.

Bolin, Bert (1925–2007)

Swedish meteorologist who served as the first chairman of the Intergovernmental
Panel on Climate Change (IPCC).

Boltwood, Bertram (1870–1957)

American radiochemist who established that lead is the final product in the radioactive decay of uranium. In 1907, he became the first to measure the age of
rocks by the decay of uranium to lead.

Bonney, Thomas George (1833–1923)

English geologist whose study of kimberlites indicated that the large diamonds had
been shattered throughout the rocks during deep explosions.

Boussingault, Jean-Baptiste (1802–1887)

Adventurous French chemist who measured the composition of gases released from
volcanoes in the Andes. He suggested that thermal instability of carbonates at
depth produced reduced carbon.

Bowen, Norman Levi (1887–1956)

Canadian geologist who revolutionized experimental petrology. Bowen's reaction
series is vital to our understanding of mineral crystallization.

Bridgman, Percy Williams (1882–1961)

American physicist who received the 1946 Nobel Prize in Physics for inventing
high-pressure apparatus and pioneering the investigation of the physics of
materials at high pressure.

Brongniart, Adolphe-Théodore (1801–1876)

French botanist whose pioneering work on extinct plants allowed him to chart
stratigraphically the history of plant life using fossils from the Carboniferous
coal formation and later formations.

Brongniart, Alexandre (1770–1847)

French chemist, mineralogist and zoologist who used characteristic fossils and the
principle of succession to understand the geological column, presenting temporal narratives of distinct environmental periods.

Brunhes, Bernard (1867–1910)

French geophysicist noted for his pioneering work on paleomagnetism. In 1905, he discovered geomagnetic reversals in surface rocks of the volcanic Massif Central, France.

Bullard, Edward (1907–1980)

In 1965, Bullard and colleagues used a computer-aided approach to provide a near perfect fit of the continents bordering the Atlantic, which provided pivotal support for the emergence of plate tectonic theory.

Cavendish, Henry (1731–1810)

British natural philosopher who conducted research into the composition of atmospheric air and the properties of different gases. He also measured the density of Earth to a value within 1.2 percent of the current accepted value.

Chamberlin, Thomas (1843–1928)

Founder of the *Journal of Geology*. He had a keen interest in glaciology and climate change, contributing much to the understanding of fluctuating carbon dioxide levels in the atmosphere.

Cloëz, Stanislas (1817–1883)

French analytical chemist who, following the Orgueil meteorite fall, was the first to examine a carbonaceous chondrite.

Coes Jr., Loring (1915–1978)

American analytical chemist at the Norton company who synthetically produced coesite, a high-pressure modification of quartz.

Cohen, Emil Wilhelm (1842–1905)

German mineralogist whose assessment of the geomorphology of the Kimberley dry diggings led him to conclude that the South African diamonds and xenoliths were carried to the surface in pipes during past eruptions.

Cox, Allan V. (1926–1987)

American geophysicist whose careful examination of rock magnetic data from different strata in the Snake River Plain disproved the hypothesis of rocks self-reversing their magnetic fields. His work on dating geomagnetic reversals, made with Richard Doell and Brent Dalrymple, was a major contribution to the theory of plate tectonics.

Craig, Harmon (1926–2003)

American geochemist, he, together with Roger Revelle and Hans Suess, made quantitative estimates of the absorption and exchange rates of carbon dioxide isotopes between the atmosphere and the oceans.

Creer, Kenneth M. (1925–2014)

British geologist who found large discrepancies between the direction of the present magnetic field and those recorded in much older rocks. This showed that continents had shifted relative to the geomagnetic pole and provided the first physical evidence of continental drift.

Curie, Marie (1867–1934)

Polish-born French physicist and chemist who developed the theory of radioactivity and techniques for isolating radioactive isotopes. She discovered the chemical elements polonium and radium.

Curie, Pierre (1859–1906)

French physicist who conducted pioneering research in crystallography, magnetism and radioactivity. He discovered the effect of temperature on paramagnetism, known as Curie's law.

Cuvier, Georges (1769–1832)

French naturalist and zoologist, best known for his work on vertebrate paleontology, particularly establishing the fields of comparative anatomy and paleontology through his work in comparing living animals with fossils.

Darwin, Charles (1809–1882)

English naturalist, geologist and biologist famous for his contribution to the science of evolution, largely summarized in his *On the Origin of Species*.

Daubrée, Gabriel Auguste (1814–1896)

French geologist who put together a consortium to study the properties of the Orgueil meteorite, classifying it along with Alais as a carbonaceous chondrite.

Davy, Humphrey (1778–1829)

English chemist who made important contributions to our understanding of volatile cycling, highlighting that human existence depends on the equilibrium of gases in the atmosphere.

de Saussure, Nicolas-Théodore (1767–1845)

Swiss chemist and plant physiologist who outlined the chemical transformations that take place during photosynthesis.

Dempster, Arthur (1886–1950)

Canadian–American physicist who is best known for his work in mass spectrometry, which led to his discovery of ^{235}U in 1935. The ^{235}U decay scheme ends in ^{207}Pb, providing a shorter-term chronometer for dating minerals.

Dietrich, Günther (1911–1972)

German oceanographer in charge of echo soundings on a *Meteor* expedition. He published a cross-section of the crest of the Mid-Atlantic Ridge.

Dietz, Robert S. (1914–1995)

Marine geologist, geophysicist and oceanographer who articulated the concept of seafloor spreading from mid-ocean ridges. He correctly suggested that convection currents move the brittle lithosphere on top of the plastic asthenosphere and that fracture zones are the result of uneven convection in the mantle.

Doell, Richard (1923–2008)

American geophysicist known for developing the timescale for geomagnetic reversals with Allan Cox and Brent Dalrymple.

du Toit, Alexander (1878–1948)

South African geologist who supported continental drift theory based on his comparisons of the geological and fossil stratigraphy from the coastlines of the Americas and Africa.

Dunn, Edward John (1844–1937)

Australian geologist who surveyed the South African diamond fields and concluded that the diamonds and xenoliths had been carried to the surface in pipes during past eruptions.

Ébelmen, Jacques-Joseph (1814–1852)

French chemist and mineralogist who made pioneering discoveries in the study of the chemical weathering of silicate minerals, making links to the deep carbon cycle.

Elsasser, Walter M. (1904–1991)

German-born American physicist who is considered to be the father of the self-sustaining dynamo theory for Earth's magnetism. His use of vector calculus seamlessly linked two distinct branches of mathematical physics: hydrodynamics and electromagnetism.

Eratosthenes (276–194 BCE)

Greek philosopher, mathematician, geographer and astronomer from Cyrene. He was the first person to calculate the circumference of Earth.

Ewing, Maurice (1906–1974)

American geophysicist and oceanographer with a particular interest in seismic reflection and refraction in ocean basins. The first director of the Lamont Geological Observatory, he made extensive oceanographic measurements throughout his career.

Fenner, Clarence Norman (1870–1949)

American geologist who, with C. S. Piggot, made the first calculations of a mineral's age using lead isotopes.

Geikie, Archibald (1839–1915)

Scottish geologist with interests in erosion and volcanic activity. Appointed first director of the Scottish branch of the Geological Survey, he voiced concerns over the assumptions Kelvin had made to calculate the age of Earth.

Gilbert, William (1544–1603)

English physician and natural philosopher who made the first advances in understanding geomagnetism and is credited with introducing the term "electricity." In *De Magnete*, he speculated that the solid Earth behaves as a giant bar magnet.

Goldschmidt, Victor (1888–1947)

Norwegian mineralogist acclaimed as one of the founding fathers of geochemistry. He developed the Goldschmidt classification, which groups elements according to five preferred host phases.

Graham, John Warren (1918–1971)

American geophysicist who developed a spinner magnetometer with which he demonstrated that sedimentary rocks retain their direction of magnetism through geological time.

Greenough, George (1778–1855)

English geologist who was the founding president of the Geological Society of London. His 1820 geological map of Britain is largely copied from William Smith's map.

Gutenberg, Beno (1889–1960)

German–American seismologist who discovered the discontinuity in seismic waves marking the boundary between the mantle and core.

Hall, Howard Tracy (1919–2008)

American physical chemist who was the first person to grow a synthetic diamond by a reproducible, verifiable and witnessed process.

Halley, Edmond (1656–1741)

An English polymath who was the first to survey and catalog the stars in the Southern Hemisphere. His pioneering observations of magnetic variation on his sea voyages made a huge contribution to the development of geomagnetism.

Hamilton, William (1730–1803)

British Ambassador to the Kingdom of Naples and keen volcanologist who made an impact through his studies on the eruptions of Vesuvius and Mount Etna, for which he was elected a Fellow of the Royal Society.

Heezen, Bruce C. (1924–1977)

Worked with supervisor Maurice Ewing and oceanographic cartographer Marie Tharp to produce a topographical map of the Atlantic Ocean basin.

Henrici, Arthur T. (1889–1943)

American bacteriologist who studied microbes in their natural environment, becoming one of the first to establish the concept of biofilms.

Hess, Harry (1906–1969)

American geologist and naval officer and one of the key players in the emergence of a unifying theory of plate tectonics. An indefatigable hands-on oceanographer, he suggested that convection in Earth's mantle could be the driving force behind the process.

Högbom, Arvid-Gustave (1857–1940)

Swedish geologist with a strong interest in the global carbon cycle. He estimated that the flux of carbon dioxide from the industrial burning of coal was comparable to that from natural geochemical sources.

Holmes, Arthur (1890–1965)

British geologist who was a pioneer of geochronology, performing the first accurate uranium–lead radiometric dating of minerals, thereby improving estimates of the age of Earth. He suggested that convection cells in the mantle were the driver moving the crust at the surface.

Hoyle, Fred (1915–2001)

British theoretical astronomer and a founder of the theory of stellar nucleosynthesis. Famous for discovering the nuclear pathway in the stellar interior for the synthesis of carbon from three alpha particles.

Humboldt, Alexander von (1769–1859)

Prussian polymath, naturalist and explorer who perceived nature as an interconnected global force. In 1793, he published his first botanical monograph on the subterranean vegetation he had observed and cataloged in the gold mines of Saxony.

Hutton, James (1726–1797)

Scottish geologist, physician, chemist and naturalist whose field studies led him to believe that Earth was perpetually being formed and recycled: this is the fundamental principle in geology known as uniformitarianism.

Irving, Edward (1927–2014)

British geologist who noticed large discrepancies between the direction of the present magnetic field and those recorded in much older rocks. He showed that continents had shifted relative to the geomagnetic pole, which provided some of the first physical evidence of continental drift.

Jannasch, Holger W. (1927–1998)

German marine microbiologist who identified that chemosynthetic bacteria absorb hydrogen sulfide and oxidize it to sulfur, sustaining life at hydrothermal vents.

Jeffreys, Harold (1891–1989)

British theoretician, geophysicist and astronomer whose analysis showed that the core of Earth could contain fluid. He strongly opposed continental drift theory because he felt the crust was too rigid for continental motion to be possible.

Joly, John (1857–1933)

Irish physicist and geologist who suggested that radioactive decay might release sufficient thermal energy to soften the seafloor, allowing for the lateral motion needed for continental drift.

Kant, Immanuel (1724–1804)

German philosopher who laid out a nebular hypothesis that the solar system had condensed from a vast cloud of dust and gas. He pictured the Milky Way as a large disk of stars that had formed in a similar manner.

Keeling, Charles David (1928–2005)

American scientist who invented a carbon dioxide detector and began recording at Mauna Loa Observatory in 1958. The series of measurements he started has continued ever since.

Kellogg, Louise Helen (1959–2019)

American geophysicist with expertise in chemical geodynamics and computational geophysics who developed advanced computational box models.

Kepler, Johannes (1571–1630)

German mathematician, astronomer and astrologer. He is known for his laws on planetary motion, developing a geometrical approach to planetary dynamics. He is also noted for making estimates of the age of Earth.

Laplace, Pierre Simon (1749–1827)

French scholar noted for his important advances in mathematics, statistics, physics and astronomy. After restating and further developing the nebular hypothesis for the origin of the solar system, he was one of the first scientists to postulate the existence of black holes and to discuss the concept of gravitational collapse.

Larmor, Joseph (1857–1942)

Irish physicist and mathematician who suggested that fluid motions in the core could generate the geomagnetic field through a dynamo mechanism, contrary to Einstein's belief that geomagnetism is related to axial rotation.

Lavoisier, Antoine-Laurent (1743–1794)

French nobleman and highly influential chemist and biologist. He is noted for his discovery of the role oxygen plays in combustion, compiling the first extensive list of elements and reforming chemical nomenclature. He estimated that carbon contributed about one-tenth of the mass of marble, chalk or calcareous rocks.

Lehmann, Inge (1888–1993)

Danish seismologist and geophysicist who discovered the discontinuity that marks the boundary between the outer and inner core.

Lewis, Henry Carvill (1853–1888)

American mineralogist and geologist who gave the term "kimberlite" to diamond-bearing pipe rocks such as those found in South African. He also made clear links between alluvial and kimberlite diamonds.

Lundblad, Erik G. (1925–2004)

Swedish engineer who helped to successfully synthesize the first microdiamonds by using a mixture of iron carbide and graphite in the sample assembly.

Lyell, Charles (1797–1859)

Scottish geologist who made significant contributions to the Earth sciences on a wide range of topics from earthquakes and volcanoes to the stratigraphy and division of the Tertiary period. He coined the terms "Paleozoic," "Mesozoic" and "Cainozoic."

Mason, Ronald G. (1916–2009)

British geophysicist and oceanographer who used towed magnetometers to map the magnetic striping on the seafloor off the coast of North America. His map furnished evidence for continental drift and plate tectonics.

Matthews, Drummond (1931–1997)

British marine geologist and geophysicist who ran magnetic survey lines between Africa and India. Together with Frederick Vine, he linked seafloor magnetic stripes with rifting and geomagnetic pole reversals.

Matuyama, Motonori (1884–1958)

Japanese geophysicist who produced a series of papers on paleomagnetic reversals, finding in 1929 a clear correlation between polarity and stratigraphic position.

Mercanton, Paul-Louis (1876–1963)

Swiss meteorologist, glaciologist and Arctic explorer who curated lava specimens from across the globe from 1910 to 1932 that confirmed the theory of magnetic reversals.

Mohorovičić, Andrija (1857–1936)

Croatian meteorologist and seismologist who discovered the boundary layer (the Moho) between the crust and mantle.

Morley, Lawrence Whittaker (1920–2013)

Canadian geophysicist best known for his studies of the magnetic properties of the seafloor. Independently of Frederick Vine and Drummond Matthews, he linked seafloor magnetic stripes with rifting and subduction processes.

Murray, John (1841–1914)

Canadian-born oceanographer and marine biologist who did most of his work in Britain. He was responsibe for the investigation of the natural history of the seafloor on the *Challenger* expedition. He took on the onerous editorial task of compiling all of the findings of the voyage.

Oldham, Richard (1858–1936)

British geologist who was the first to clearly identify the separate arrivals of different seismic waves from the same earthquake. His analysis of travel times of seismic waves from the 1897 earthquake in Assam, India, led him to suggest that Earth has a dense core.

Oliver, John "Jack" Ertle (1923–2011)

American geophysicist known for working with Bryan L. Isacks and Lynn R. Sykes to add seismological data to models of plate tectonics. Their global plot of earthquake epicenters revealed the outlines of the major lithospheric plates.

Ortelius, Abraham (1527–1598)

Flemish cartographer and geographer who published the first modern atlas in 1570. His third edition suggested that the Americas may have once been attached to Europe and Africa but then was wrenched away by earthquakes and floods.

Patterson, Clair Cameron (1922–1995)

American geochemist who, in collaboration with George Tilton, developed lead–lead dating, calculating the age of Earth at 4.55 billion years.

Perry, John (1850–1920)

Irish mathematician and engineer, his publications challenged Kelvin's assumption that Earth's interior has low thermal conductivity and is structurally homogeneous.

Pigott, Charles S. (1892–1973)

Pioneer of ocean-bottom marine research who obtained reliable dates for sediments from the northern Atlantic Ocean and Caribbean Basin.

Revelle, Roger (1909–1991)

American oceanographer who was instrumental in enabling Ronald Mason to map the magnetism of the seafloor. He was also one of the first to study anthropogenic global warming and, together with Harmon Craig and Hans Suess, he made quantitative estimates of the absorption and exchange rates of CO_2 isotopes between the atmosphere and the oceans.

Rhodes, Cecil John (1853–1902)

Victorian mining magnate who consolidated the diamond mines of South Africa under the name De Beers. At one point, he controlled nine-tenths of the world trade in diamond gemstones.

Runcorn, Keith (1922–1995)

British geophysicist who discovered that each continent has a distinctive polar wander path. Through paleomagnetic reconstructions of the relative motions of Europe and America, he revived the controversial theory of continental drift.

Rutherford, Ernest (1871–1937)

New Zealand-born British physicist who laid the foundations of nuclear physics. He and Frederick Soddy were the first to use the concept of the radioactive half-life.

Sabine, Edward (1788–1883)

Irish astronomer, geophysicist and explorer who led the effort to establish a system of magnetic observatories in various parts of the British Empire.

Sedgwick, Adam (1785–1873)

British geologist who became highly influential as President of the Geological Society of London from 1829 to 1831, particularly concerning his acceptance of uniformitarianism. He proposed the Cambrian, Devonian and Silurian periods of the geological timescale, and he tutored Darwin's studies of field geology.

Smith, William (1769–1839)

English land surveyor whose observations of fossils within different strata contributed to the foundations of stratigraphy. Famously, he created the first nationwide geological maps.

Spallanzani, Lazarro (1729–1799)

Italian cleric, biologist and physiologist who advanced the idea that the release of water or carbonic acid from minerals under pressure was sustaining volcanic gas emissions.

Steno, Nicholas (1638–1686)

Danish natural philosopher who made important contributions to anatomy and Earth sciences. He proposed four laws of stratigraphy that are still used today.

Story-Maskelyne, Nevil (1823–1911)

English geologist and keeper of minerals at the British Museum. Along with museum colleagues, he determined that diamondiferous rocks from South Africa were igneous in their origin.

Strachey, John (1671–1743)

English geologist and topographer, his publications displayed for the first time the order and regularity of strata. This set the standard from which modern stratigraphic cross-sections developed.

Strutt, Robert (1875–1947); Fourth Baron Rayleigh

British physicist who made minimum estimates of the age of thorianite and uraninite, revealing the extreme antiquity of specific minerals and the process of mineral evolution.

Suess, Eduard (1831–1914)

Austrian geographer and geologist who hypothesized the former existence of the supercontinent Gondwana. He also helped to develop the fixist theory, suggesting that parts of paleocontinents had sundered to form ocean bottoms.

Suess, Han (1909–1993)

Austrian-born American physical chemist and nuclear physicist who made quantitative estimates of the absorption and exchange rates of CO_2 isotopes between the atmosphere and the oceans.

Tharp, Marie (1920–2006)

American oceanographic cartographer who worked with Bruce Heezen and Maurice Ewing to produce the first topographic maps of the Atlantic Ocean basin, revealing the continuous mid-ocean ridge and its central rift.

Thénard, Louis Jacques (1777–1857)

French chemist who analyzed the chemical composition of the Alais meteorite, the first carbonaceous chondrite to be discovered.

Thomson, Charles Wyville (1830–1882)

Scottish natural historian and marine zoologist who served as the chief scientist on the *Challenger* expedition that set sail in 1872 on a research voyage that was transformative for oceanography.

Thomson, William (1824–1907); Baron Kelvin

Scots–Irish mathematical physicist and engineer, known for his work in classical thermodynamics, which he applied to estimate the age of the Sun and thus deduce an age for Earth.

Tyndall, John (1820–1893)

Irish physicist noted for his studies on diamagnetism and infrared radiation. He found that atmospheric carbon dioxide and methane are effective at trapping infrared radiation.

Urey, Harold (1893–1981)

American physical chemist whose pioneering work on isotopes of carbon resulted in the Ébelmen–Urey reaction, the key to understanding the silicate–carbonate subcycle.

Ussher, James (1581–1656)

Church of Ireland Archbishop of Armagh from 1625 to 1656, remembered for his chronology of Creation (4004 BCE), derived from a literal interpretation of the Bible.

Vacquier, Victor (1907–2009)

Russian-born American geophysicist and oceanographer who invented the fluxgate magnetometer. His discovery of large shifts in patterns in the Mendocino fracture zone was a major impetus behind the theory of plate tectonics.

van Allen, James (1914–2006)

American space scientist who was instrumental in establishing the field of magnetospheric research in space. He organized the International Geophysical Year (1957–1958).

Vanuxem, Lardner (1792–1848)

American geologist who recorded the presence of igneous dykes similar to kimberlites in multiple locations in the USA.

Wegener, Alfred (1880–1930)

German polar researcher, geophysicist and meteorologist. A daring pioneer of polar science research, he is mainly remembered for proposing the theory of continental drift.

Weichert, Emil (1861–1928)

German geophysicist who was the first to present a verifiable model of a layered structure of Earth.

Wilson, John Tuzo (1908–1993)

Canadian geologist and geophysicist who was the first to develop the idea of transform faults. The Wilson cycle of continental divergence and collision bears his name.

Yoder Jr., Hatten Schuyler (1921–2003)

American geophysicist and petrologist who worked at the Carnegie Geophysical Laboratory, where he conducted high-pressure, high-temperature experiments on igneous rocks and minerals.

ZoBell, Claude E. (1904–1989)

American bacteriologist who studied microbes in their natural environment and promoted the concept of biofilms.

Index